Graph-Based Semi-Supervised Learning

Synthesis Lectures on Artificial Intelligence and Machine Learning

Editors
Ronald J. Brachman, *Yahoo! Labs*
William W. Cohen, *Carnegie Mellon University*
Peter Stone, *University of Texas at Austin*

Learning with Support Vector Machines
Colin Campbell and Yiming Ying
2011

Algorithms for Reinforcement Learning
Csaba Szepesvári
2010

Data Integration: The Relational Logic Approach
Michael Genesereth
2010

Markov Logic: An Interface Layer for Artificial Intelligence
Pedro Domingos and Daniel Lowd
2009

Introduction to Semi-Supervised Learning
XiaojinZhu and Andrew B.Goldberg
2009

Action Programming Languages
Michael Thielscher
2008

Representation Discovery using Harmonic Analysis
Sridhar Mahadevan
2008

Essentials of Game Theory: A Concise Multidisciplinary Introduction
Kevin Leyton-Brown and Yoav Shoham
2008

A Concise Introduction to Multiagent Systems and Distributed Artificial Intelligence
Nikos Vlassis
2007

Intelligent Autonomous Robotics: A Robot Soccer Case Study
Peter Stone
2007

Graph-Based Semi-Supervised Learning

Amarnag Subramanya and Partha Pratim Talukdar

ISBN: 978-3-031-00443-8   paperback
ISBN: 978-3-031-01571-7   ebook

DOI 10.1007/978-3-031-01571-7

A Publication in the Springer series
SYNTHESIS LECTURES ON ARTIFICIAL INTELLIGENCE AND MACHINE LEARNING

Lecture #29
Series Editors: Ronald J. Brachman, *Yahoo! Labs*
            William W. Cohen, *Carnegie Mellon University*
            Peter Stone, *University of Texas at Austin*
Series ISSN
Print 1939-4608   Electronic 1939-4616

Graph-Based Semi-Supervised Learning

Amarnag Subramanya
Google Research, Mountain View, USA

Partha Pratim Talukdar
Indian Institute of Science, Bangalore, India

SYNTHESIS LECTURES ON ARTIFICIAL INTELLIGENCE AND MACHINE LEARNING #29

ABSTRACT

While labeled data is expensive to prepare, ever increasing amounts of unlabeled data is becoming widely available. In order to adapt to this phenomenon, several semi-supervised learning (SSL) algorithms, which learn from labeled as well as unlabeled data, have been developed. In a separate line of work, researchers have started to realize that graphs provide a natural way to represent data in a variety of domains. Graph-based SSL algorithms, which bring together these two lines of work, have been shown to outperform the state-of-the-art in many applications in speech processing, computer vision, natural language processing, and other areas of Artificial Intelligence. Recognizing this promising and emerging area of research, this synthesis lecture focuses on graph-based SSL algorithms (e.g., label propagation methods). Our hope is that after reading this book, the reader will walk away with the following: (1) an in-depth knowledge of the current state-of-the-art in graph-based SSL algorithms, and the ability to implement them; (2) the ability to decide on the suitability of graph-based SSL methods for a problem; and (3) familiarity with different applications where graph-based SSL methods have been successfully applied.

KEYWORDS

semi-supervised learning, graph-based semi-supervised learning, manifold learning, graph-based learning, transductive learning, inductive learning, nonparametric learning, graph Laplacian, label propagation, scalable machine learning

To the loving memory of my father

A.S.

To the loving memory of Bamoni, my beloved sister

P.P.T.

Contents

CHAPTER 1

Introduction

Learning from limited amounts of labeled data, also referred to as *supervised learning*, is the most popular paradigm in machine learning. This approach to machine learning has been very successful with applications ranging from spam email classification to optical character recognition to speech recognition. Such labeled data is usually prepared based on human inputs, which is expensive to obtain, difficult to scale, and often error prone. At the same time, unlabeled data is readily available in large quantities in many domains. In order to benefit from such widely available unlabeled data, several *semi-supervised learning (SSL) algorithms* have been developed over the years [Zhu and Goldberg, 2009]. SSL algorithms thus benefit from both labeled as well as unlabeled data.

With the explosive growth of the World Wide Web, graph structured datasets are becoming widespread, e.g., online social networks, hyperlinked web graph, user-product graph, user-video watched graph, etc. In a separate line of work, researchers have started to realize that graphs provide a natural way to represent data in a variety of other domains. In such datasets, nodes correspond to data instances, while edges represent relationships among nodes (instances).

Graph-based SSL techniques bring together these two lines of research. In particular, starting with the graph structure and label information about a subset of the nodes, graph SSL algorithms classify the remainder of the nodes in the graph. Graph-based SSL algorithms have been shown to outperform *non* graph-based SSL approaches [Subramanya and Bilmes, 2010]. Furthermore, majority of the graph-based SSL approaches can be optimized using convex optimization techniques and are both easily scalable and parallelizable. Graph-based SSL algorithms have been successfully used in applications as diverse as phone classification [Subramanya and Bilmes, 2010], part-of-speech (POS) tagging [Subramanya et al., 2010], statistical machine translation (SMT) [Alexandrescu and Kirchhoff, 2009], word sense disambiguation, semantic parsing [Das and Smith, 2012], knowledge acquisition [Wijaya et al., 2013], sentiment analysis in social media [Lerman et al., 2009], text categorization [Subramanya and Bilmes, 2008], and many others. Recognizing this promising and emerging area of research, this book provides an introduction to graph-based SSL techniques.

To give a concrete example, let us consider the graph shown in Figure 1.1. This graph consists of six nodes, where edge weight represents the degree of similarity between the two nodes connected by the edge. The nodes *Microsoft* and *Bangalore* are known to have the labels *Company* and *City*, respectively. Starting from this setting, a graph-based SSL algorithm aims to classify the remaining initially unlabeled four nodes.

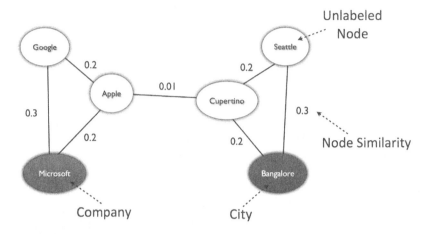

Figure 1.1: Starting with the setting shown above, a graph-based SSL may be used to classify the four unlabeled nodes. Ideally, we would like to infer that *Seattle* and *Cupertino* are both *Cities*, while *Google* and *Apple* are *Companies*.

In order to understand Graph-based SSL techniques, it might be informative to first understand their position in the ontology of learning algorithms. To this end, we first present some conceptual definitions, and motivations behind different learning settings.

Definition 1.1 Instance. An *instance* $\mathbf{x}$ represents a specific object. The instance is often represented by a d-dimensional *feature vector* $\mathbf{x} = [x_1, \ldots x_d] \in \mathcal{R}^{d \times 1}$, where each dimension is commonly referred to as feature. Each instance may be associated with one or more labels from the label set $\mathcal{Y}$. An instance with known label assignment is called a *labeled (or training) instance*, and otherwise called an *unlabeled instance*.

Based on the amount of supervisory information in the training data, machine learning can be categorized into *unsupervised learning*, *supervised learning*, and *semi-supervised learning*.

1.1 UNSUPERVISED LEARNING

The training data in the case of unsupervised learning contains *no supervisory* information. Here, $\mathcal{D} = \{\mathbf{x}_i\}_{i=1}^{n} = \{\mathbf{x}_1, \ldots, \mathbf{x}_n\}$. It is assumed that each $\mathbf{x}_i$ is sampled independently from a distribution $P(\mathbf{x}), \mathbf{x} \in \mathcal{X}$. Thus, these samples are independent and identically distributed or *i.i.d.* Note that the underlying distribution $P(\mathbf{x})$ is not revealed to the learning algorithm.

Examples of unsupervised learning problems include, clustering, dimensionality reduction, and density estimation. The goal of clustering is to group (cluster) similar instances in $\mathcal{D}$, while dimensionality reduction aims to represent each sample with a lower dimensional feature vector with as little loss of information as possible. Density estimation is the problem of estimating of the parameters of the underlying distribution that generated the $\mathcal{D}$.

1.2   SUPERVISED LEARNING

In supervised learning, every sample in $\mathcal{D}$ has both the input $\mathbf{x} \in \mathcal{X}$ and the corresponding expected response $y \in \mathcal{Y}$. Here, $\mathcal{Y}$ represents the set of possible outputs. The expected response is also referred to as the *label*. Formally, given a training set $\mathcal{D} = \{(\mathbf{x}_i, y_i)\}_{i=1}^{n}$, the goal of a supervised learning algorithm is to train a function $f : \mathcal{X} \to \mathcal{Y}$ where $f \in \mathcal{F}$. Here, $\mathcal{F}$ is some pre-defined family of functions. Each training sample $(\mathbf{x}_i, y_i)$ is assumed to be sampled independently from a joint distribution $P(\mathbf{x}, y), \mathbf{x} \in \mathcal{X}, y \in \mathcal{Y}$, which is not revealed to the learning algorithm. Once we have a trained model, $f^*(\mathbf{x})$, given a new input $\mathbf{x}' \in \mathcal{X}$ we can predict the label $\hat{y} = f^*(\mathbf{x}')$.

When one uses probabilistic models, the decision function $f(x)$ is given by $p(\mathbb{Y}|\mathbb{X}; \Theta)$.[1] Here, $\mathbb{Y}$ and $\mathbb{X}$ are random variables defined over the output and input domains respectively and $p(\mathbb{Y}|\mathbb{X}; \Theta)$ is a conditional distribution parameterized by $\Theta \in \mathcal{R}^{|\Theta|}$. Thus, training in this case involves learning Θ given $\mathcal{D}$. Once we have a trained model Θ^*, given a new input $\mathbf{x}' \in \mathcal{X}$, the most likely class can be computed by $y^* = \text{argmax}_{y \in \mathcal{Y}}\, p(y|\mathbb{X} = \mathbf{x}'; \Theta^*)$.

Based on the nature of the output domain $\mathcal{Y}$, supervised learning is categorized into classification and regression. Supervised learning problems where $\mathcal{Y}$ is discrete are referred to as *classification* while those in which $\mathcal{Y}$ is continuous are called *regression*. In the classification case, each $y \in \mathcal{Y}$ is referred to as a *class*. Problems with only two classes, i.e., $|\mathcal{Y}| = 2$, are called *binary classification* while *multi-class* problems have $|\mathcal{Y}| > 2$. In the case of binary classification, $y \in \mathcal{R}$ can be used to represent the result of classification (or annotation) but in the multi-class setting $y \in \mathcal{R}_+^{|\mathcal{Y}|}$. In other words, the size of y is equal to the number of classes. The values may either represent a score indicating how likely it is for a particular input to belong to a particular class or may be a valid probability distribution.

Examples of supervised learning algorithms include, support vector machines (SVM) [Scholkopf and Smola, 2001] and maximum entropy classifiers [Mitchell, 1997]. Supervised learning is perhaps one of the most well researched areas of machine learning and a large number of machine learning applications are built using supervised learning [Duda et al., 2001, Kotsiantis, 2007, Turk and Pentland, 1991].

Depending on the complexity of the function class $\mathcal{F}$, the number of samples required to accurately learn the mapping, f, can sometimes be extremely large. Training with insufficient amounts of data can lead to poor performance on unseen data. This is a drawback of supervised learning as annotating large amounts of data requires extensive human supervisory efforts. This can be both time consuming and often error prone. Unsupervised learning, on the other hand, requires no

---

[1]While discriminative models directly model this distribution, in the case of generative models, the Bayes rule is used to compute the class conditional distribution $p(y|x)$. For more details, see Mitchell [1997].

"labeled" training data but suffers from the inability for one to specify the expected output for a given input.

1.3   SEMI-SUPERVISED LEARNING (SSL)

Semi-supervised learning (SSL) lies somewhere between supervised and unsupervised learning. It combines the positives of both supervised and unsupervised learning. Here only a small amount of the training set $\mathcal{D}$ is labeled while a relatively large fraction of the training data is left unlabeled. The goal of a SSL algorithm is to learn a function $f : \mathcal{X} \rightarrow \mathcal{Y}, f \in \mathcal{F}$ given a training set $\mathcal{D} = \{\mathcal{D}_l, \mathcal{D}_u\}$ where $\mathcal{D}_l = \{(\mathbf{x}_i, y_i)\}_{i=1}^{n_l}$ represents the labeled portion of the training data while $\mathcal{D}_u = \{\mathbf{x}_i\}_{i=1}^{n_u}$ are the unlabeled samples. Thus, we have n_l labeled examples and n_u unlabeled examples. We assume that the total number of training samples $n \triangleq n_l + n_u$. In most practical settings, $n_u \gg n_l$. Each labeled training sample $(\mathbf{x}_i, y_i)$ is assumed to be sampled independently from a joint distribution $P(\mathbf{x}, y), \mathbf{x} \in \mathcal{X}, y \in \mathcal{Y}$ while the unlabeled examples are sampled from $P(\mathbf{x}) = \sum_y P(\mathbf{x}, y)$. Like in the cases above, these distributions are not revealed to the learning algorithm. Note that like in the case of supervised learning, based on the type of the output space $\mathcal{Y}$, we have *semi-supervised classification* and *semi-supervised regression*. In this book we will be focussing on semi-supervised classification. Unless stated otherwise, SSL implies semi-supervised classification.

How does unlabeled data help?   The goal of an SSL algorithm is to learn the mapping $f : \mathcal{X} \rightarrow \mathcal{Y}$. However, it appears that the unlabeled data contains no information about this mapping. In general, SSL algorithms make one or more of the following assumptions so that information available in the unlabeled data can influence $f : \mathcal{X} \rightarrow \mathcal{Y}$.

1. *Smoothness Assumption*—if two points in a high-density region are close then their corresponding outputs are also close. In regression problems, for example, the above assumption implies that the function $f : \mathcal{X} \rightarrow \mathcal{Y}$ is continuous.

2. *Cluster Assumption*—if two points are in the same cluster, they are likely to be of the same class. Another way to state this assumption would be to say that the decision boundary should lie in a low-density region. Transductive SVMs and some of the graph-based SSL algorithms are based on this assumption.

3. *Manifold Assumption*—high-dimensional data lies within a low-dimensional manifold. This is very important owing to the fact that most machine learning algorithms suffer from the "curse of dimensionality." Thus, being able to handle the data on a relatively low-dimensional manifold can often be very advantageous for the algorithms.

One of the earliest SSL algorithm is *self-training* (or self-learning or self-labeling) [Scudder, 1965]. In many instances, *expectation-maximization* (EM) [Dempster et al., 1977] can also be seen as an SSL algorithm. EM is a general procedure to maximize the likelihood of the data given a model with hidden variables and is guaranteed to converge to a local maxima. EM lends itself naturally to SSL as the labels for the unlabeled data can be treated as missing (hidden) variables. Example of algorithms that use EM for SSL include [Hosmer, 1973, McLachlan and Ganesalingam, 1982, Nigam, 2001]. Co-training is another SSL algorithm [Blum and Mitchell, 1998] that is related to self-training and takes advantage of multiple views of the data. Transductive support vector machines (TSVM) [Vapnik, 1998] are based on the premise that the decision boundary must avoid high density regions in the input space. They are related to the fully supervised support vector machines. For more details on SSL algorithms, see [Chapelle et al., 2007].

1.4   GRAPH-BASED SEMI-SUPERVISED LEARNING

Graph-based SSL algorithms are an important sub-class of SSL techniques that have received much attention in the recent past [Chapelle et al., 2007, Zhu, 2005]. Here, one assumes that the data (both labeled and unlabeled) is embedded within a low-dimensional manifold that may be reasonably expressed by a graph. Each data sample is represented by a vertex in a weighted graph with the weights providing a measure of similarity between vertices. Thus, taking a graph-based approach to solving a SSL problem involves the following steps:

1. graph construction (if no input graph exists),

2. injecting seed labels on a subset of the nodes, and

3. inferring labels on unlabeled nodes in the graph.

Graph construction will be the topic of Chapter 2 while inference algorithms will be discussed in Chapter 3. There are a number of reasons why graph-based SSL algorithms are very attractive.

1. *Graphs are everywhere*: As stated earlier, many data sets of current interest are naturally represented by graphs. For example, the Web is a hyperlinked graph, social network is a graph, communication networks are graphs, etc.

2. *Convexity*: As we will see in Chapter 3, majority of the graph-based SSL algorithms involve optimizing a convex objective.

3. *Ease of scalability*: As SSL is based on the premise that large amounts of unlabeled data improve performance, it is very important for SSL algorithms to scale. This is crucial in many application domains (see Chapter 5) where ability to handle large datasets is a prerequisite. As we will see in Chapter 4, compared to other (non-graph-based) SSL algorithms, many graph-based SSL techniques can be easily parallelized. For example, state-of-the-art TSVMs can only handle tens of thousands of samples when using an arbitrary kernel; Karlen et al. [2008] report that for a problem with about 70,000 samples (both labeled and unlabeled included), it took about 42 h to train a TSVM.

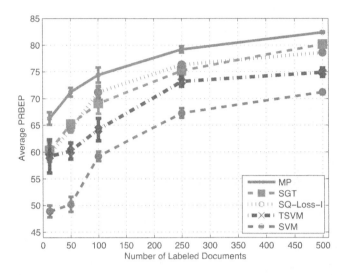

Figure 1.2: Performance of various SSL techniques on a text categorization problem. The y-axis represents average precision-recall break even point (PRBEP) and so larger is better. Measure propagation (MP), spectral graph transduction (SGT), and SQ-Loss-I are all graph-based SSL algorithms. SVM is fully supervised and TSVM is a SSL algorithm that does not make use of a graph. It can be seen that the graph-based approaches outperform other SSL and supervised techniques.

4. *Effective in practice*: Finally, graph SSL algorithms tend to be effective in practice. For example, Figure 1.2 shows a comparison of the performance of a number of SSL algorithms on a text categorization task [Subramanya and Bilmes, 2010]. The input here is a document (e.g., news article) and the goal is to classify it into a one of many classes (e.g., sport, politics). It can be seen that the graph-based SSL algorithms (MP, SGT, and SQ-Loss-I) outperform the SSL approaches and supervised techniques.

Next, we define a number of terms that we will use in this book.

Definition 1.2   Graph.   A graph is an ordered pair, $G = (V, E)$ where $V = \{1, \ldots, |V|\}$ is the set of vertices (or nodes) and $E \subseteq \{V \times V\}$ is the set of edges.

Definition 1.3   Directed and Undirected Graphs.   A graph in which the edges have no orientation is called an undirected graph. In the case of a directed graph the edges have a direction associated with them. Thus, the set of edges E in a directed graph consists of ordered pairs of vertices.

As explained above, graph-based SSL algorithms start by representing the data (labeled and unlabeled) as a graph. We assume that vertex $i \in V$ represents input sample $\mathbf{x}_i$. We will be using both i and $\mathbf{x}_i$ to refer to the ith vertex in the graph.

Definition 1.4 Weighted Graph.   A graph is said to be weighted if there is a number or weight associated with every edge in the graph. Given an edge (i, j), where $i, j \in V$, we use W_{ij} to denote the weight of the edge. We will use the matrix $W \in \mathcal{R}^{n \times n}$ to denote all the edge weights, i.e., $W_{ij} = [W]_{ij}$. Also, $G = (V, E, W)$ represents a weighted graph.

While, in general, there are no constraints on W_{ij}, for all graphs in this book we assume that $W_{ij} \geq 0$. Further we assume that $W_{ij} = 0$ if and only if there is no edge between vertices i and j. Also in the case of an undirected graph $W_{ij} = W_{ji}$ and thus W is a symmetric matrix. Majority of the graphs in this book will be weighted undirected graphs.

Definition 1.5 Connected component.   Given an undirected graph, a connected component is a subgraph in which all pairs of vertices are connected to each other by a path.

Definition 1.6 Degree of a Vertex.   The degree D_{ii} of vertex i is given by $D_{ii} = \sum_j W_{ij}$.
In the case of an unweighted graph, the degree of vertex is equal to its number of neighbors.

Definition 1.7 Unnormalized Graph Laplacian.   The unnormalized graph Laplacian is given by $L = D - W$. Here $D \in \mathcal{R}^{n \times n}$ is a diagonal matrix such that D_{ii} is the degree of node i and $D_{ij} = 0 \ \forall i \neq j$. L is a positive semi-definite matrix.

Definition 1.8 Normalized Graph Laplacian.   The normalized graph Laplacian is given by $\mathcal{L} = D^{-1/2} L D^{1/2}$.

1.4.1   INDUCTIVE VS. TRANSDUCTIVE SSL

Definition 1.9 Inductive Algorithm.   Given a training set consisting of labeled and unlabeled data, $\mathcal{D} = \{\{\mathbf{x}_i, y_i\}_{i=1}^{n_l}, \{\mathbf{x}_i\}_{i=1}^{n_u}\}$, the goal of an inductive algorithm is to learn a function $f : \mathcal{X} \to \mathcal{Y}$. Thus f is able to predict the output y for any input $\mathbf{x} \in \mathcal{X}$.

Definition 1.10 Transductive Algorithm.   Given a training set consisting of labeled and unlabeled data, $\mathcal{D} = \{\{\mathbf{x}_i, y_i\}_{i=1}^{n_l}, \{\mathbf{x}_i\}_{i=1}^{n_u}\}$, the goal of a transductive algorithm is to learn a function $f : \mathcal{X}^n \to \mathcal{Y}^n$. In other words, f only predicts the labels for the unlabeled data $\{\mathbf{x}_i\}_{i=1}^{n_u}$.

Parametric vs. Nonparametric Models

Orthogonal to these are the concepts of parametric and nonparametric models. In case of nonparametric models, the model *structure* is not specified *a priori*, and is instead allowed to change as dictated by the data. In contrast, in case of parametric models, the data is assumed to come from a type of probability distribution, and the goal is to recover the parameters of the probability distribution. Most graph SSL techniques are nonparametric in nature.

1.5   BOOK ORGANIZATION

The rest of the book is organized as follows. In Chapter 2, we look at how a graph can be constructed from the data when no graph is given as input. In Chapter 3, we look at various graph SSL inference techniques. Then in Chapter 4, we look at issues related to scaling graph SSL to large datasets. Applications to various real-world problems will be the focus of Chapter 5.

CHAPTER 2

Graph Construction

As mentioned earlier, the application of a graph-based SSL algorithm requires solving two sub-problems: (a) constructing a graph over the input data (if one is not already available) and (b) inferring the labels on the unlabeled samples in the input or estimating the model parameters. While many algorithms have been developed for label inference [Subramanya and Bilmes, 2010, Zhu et al., 2003], until recently, little attention has been paid to the crucial graph construction phase [Daitch et al., 2009, Jebara et al., 2009, Maier et al., 2009, Wang and Zhang, 2008]. We review some of the graph construction methods in this chapter. Inference algorithms will be the focus of the next chapter.

Graph-based SSL methods operate on a graph where a node represents a data instance and a pair of nodes are connected by a weighted edge. Some nodes in the graph are labeled, i.e., we know the expected output label for the inputs associated with these nodes. In some real world domains (e.g., the hyperlinked web, social networks, citation networks, etc.) the data is relational in nature and there is already an implicit underlying graph. Graph-based methods are thus a natural fit for SSL problems in these domains. However, for a majority of learning tasks, the data instances are assumed to be independent and identically distributed (i.i.d.) and as a result one needs to induce a graph to make these problems amenable to graph-based SSL methods. The graph construction techniques presented in this chapter is particularly relevant for this latter case.

Graph construction methods can be broadly classified into the following two categories.

- **Task-independent Graph Construction**: These methods do not use labeled data for graph construction, and hence they are task-independent or unsupervised in nature. k-Nearest Neighbor (k-NN) and ϵ-neighborhood are the two most commonly used graph construction methods which fall under this category.

  b-matching [Jebara et al., 2009] is another task-independent method where the goal is to induce a regular graph. A graph is said to be regular if every vertex has the same number of neighbors. In contrast, k-NN based graph construction can lead to irregular graphs.

  Daitch et al. [2009] have proposed the concept of *hard* and α-*soft* graphs. A graph is defined to be hard if each node has a weighted degree (sum of weights of the edges connected to it) of at least 1. While, in case of α-soft graphs, this degree requirement is relaxed to allow some nodes (e.g., outliers) to violate the above condition. Daitch et al. [2009] show how one can induce hard and α-soft graphs. We will also review some of the analyses of the properties of the induced graphs and their effectiveness in practical settings.

- **Task-dependent Graph Construction**: In contrast to the methods described above, algorithms that fall under this category make use of both the labeled and unlabeled data for graph construction. Labeled data can be used as a prior for adapting the graph to the task at hand. While a few task-independent construction methods exist, the area of task-dependent graph construction has received even lesser attention.

  Inference-driven Metric Learning (IDML) [Dhillon et al., 2010] is a semi-supervised metric learning algorithm which exploits labeled as well as widely available unlabeled data to learn a metric. This learned metric can in turn be used to construct a task-specific graph.

  Kernel-alignment based spectral kernel design [Zhu et al., 2007] is an example of task-dependent graph construction. The idea here is to construct a kernel for supervised learning, starting with a fixed graph and in the process transforming the spectrum of its Laplacian. Although this is not exactly a graph construction method, it provides interesting insights into the graph Laplacian and its importance for graph-based SSL, which, in turn, may be exploited for graph construction.

2.1   PROBLEM STATEMENT

Recall that the training data in the case of SSL is $\mathcal{D} = \{\mathcal{D}_l, \mathcal{D}_u\}$ where $\mathcal{D}_l = \{(\mathbf{x}_i, y_i)\}_{i=1}^{n_l}$ represents the labeled portion of the training data while $\mathcal{D}_u = \{\mathbf{x}_i\}_{i=1}^{n_u}$ are the unlabeled samples. Also recall that $n \triangleq n_l + n_u$.

The goal of graph construction is to discover a graph $G = (V, E, W)$ where V is the set of vertices, $E \subseteq V \times V$ are the edges and $W \in \mathcal{R}^{n \times n}$ are the weights on the edges. Each vertex in the graph represents an input sample and thus the number of vertices in the graph $|V| = n$. As the vertices are fixed (assuming that $\mathcal{D}$ is fixed), the task of graph construction involves estimating E and W. Unless otherwise stated, for graphs in this book we make the following assumptions.

1. The graph is undirected and so W is symmetric. In other words, $W_{ij} = w_{ji}$. All edge weights are non-negative, i.e., $W_{ij} \geq 0, \ \forall i, j$

2. $W_{ij} = 0$ implies the absence of an edge between nodes i and j. Thus, graph construction in essence involves estimating W.

3. There are no self-loops, i.e., $W_{ii} = 0, \ \forall \ 1 \leq i \leq n$.

Some of the methods discussed in this chapter may impose additional restrictions on W and we will specify them where necessary.

Definition 2.1   Distance Metric and Distance Matrix.   Let $d : \mathcal{X} \times \mathcal{X} \to \mathcal{R}_+$ be a distance metric. Given n points $\{\mathbf{x}_i\}_{i=1}^n$ where $\mathbf{x}_i \in \mathcal{X}$, let $\Delta \in \mathbb{R}_+^{n \times n}$ be the distance matrix where $\Delta_{ij} = d(\mathbf{x}_i, \mathbf{x}_j)$. As d is a metric, Δ is symmetric.

Definition 2.2 Kernel Matrix.   Given a kernel $k(\mathbf{x}_i, \mathbf{x}_j)$, and a set of n samples $\{\mathbf{x}_i\}_{i=1}^{n}$ where $\mathbf{x}_i \in \mathcal{X}$, the kernel matrix is denoted by $K \in \mathcal{R}^{n \times n}$. Here $[K]_{ij} = k(\mathbf{x}_i, \mathbf{x}_j)$. K is sometimes as also referred to as the *Gram matrix*. The kernel matrix is positive semi-definite.

2.2   TASK-INDEPENDENT GRAPH CONSTRUCTION

2.2.1 k-NEAREST NEIGHBOR (k-NN) AND ϵ-NEIGHBORHOOD METHODS

k-Nearest Neighbors (k-NN) and ϵ-neighborhood are two of the most popular methods for constructing a graph. They require a similarity ($\text{sim}(\mathbf{x}_i, \mathbf{x}_j)$) or distance function ($d(\mathbf{x}_i, \mathbf{x}_j)$) using which similarity or distance between any two points in the input data can be computed. The edge weights are given by

$$W_{ij} = \begin{cases} \text{sim}(\mathbf{x}_i, \mathbf{x}_j) & \text{if } i \text{ is the nearest neighbor of } j \text{ or vice versa,} \\ 0 & \text{otherwise.} \end{cases} \tag{2.1}$$

As we seek a symmetric W, an edge is added between nodes i and j if either i is a nearest neighbor of j or vice versa. Thus, some nodes in the graph can have a very large number of neighbors or degree leading to irregular graphs. A graph is said to be regular if every node has the same degree. For example, consider Figure 2.1 which shows the 1-NN graph of a set of five points arranged in specific configuration. It can be seen that the central node ends up with degree of 4 which is significantly higher than the degree (1) of all the other four nodes. Nearest neighbor methods for graph construction are also computationally very expensive. This is particularly a problem for large data sets. However, a variety of alternatives and approximations have been developed over the years [Bentley, 1980, Beygelzimer et al., 2006]. These will be discussed in more detail in Chapter 4.

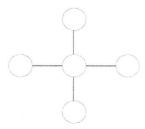

Figure 2.1:   1-NN graph for a set of five points. Note that the central node ends up with degree 4 while all other nodes have degree 1, resulting in an irregular graph.

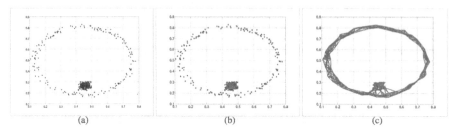

Figure 2.2: k-NN and ϵ-neighborhood graphs constructed from a synthetic dataset [Jebara et al., 2009] (a) The synthetic dataset, (b) ϵ-neighborhood graph, and (c) k-NN graph (k = 10). The ϵ-neighborhood is quite sensitive to the choice of ϵ and it may return graphs with disconnected components as in (b).

In ϵ-neighborhood-based graph construction, an undirected edge between two nodes is added if the distance between them is smaller than ϵ, where $\epsilon > 0$ is a predefined constant. In other words, given a point $\mathbf{x}$, a ball (specific to the chosen distance metric) of radius ϵ is drawn around it and undirected edges are added between $\mathbf{x}$ and all points inside this ball.

k-NN methods enjoy certain favorable properties compared to ϵ-neighborhood based graphs. For example, k-NN methods are adaptive to scale and density while an inaccurate choice of ϵ may result in disconnected graphs [Jebara et al., 2009]. An empirical verification of this phenomenon on a synthetic dataset is shown in Figure 2.2. Further, Jebara et al. [2009] showed that k-NN methods tend to perform better in practice compared to ϵ-neighborhood based methods.

2.2.2   GRAPH CONSTRUCTION USING b-MATCHING

As discussed above, k-NN graphs, contrary to their name, often lead to graphs where different nodes have different degrees. Jebara et al. [2009] proposed b-matching which guarantees that every node in the resulting graph has exactly b neighbors. Using b-matching for graph construction involves the following two steps: (a) graph sparsification and (b) edge re-weighting.

Graph Sparsification

Given a set of n points, there are total of $\binom{n}{2}$ possible (undirected) edges between these points. Creating a graph with all of these edges leads to a dense graph (clique) which not only puts significant computational burden on any subsequent inference algorithm but can also have adverse effects on the accuracy. The goal of the sparsification step is to select a subset of the above edges to be present in the final graph.

Sparsification removes edges by estimating a matrix, $P \in \{0, 1\}^{n \times n}$, where $P_{ij} = 1$ implies an edge between points $\mathbf{x}_i$ and $\mathbf{x}_j$ in the final graph while $P_{ij} = 0$ indicates the absence of an edge. There are no self-loops, i.e., $P_{ii} = 0$. k-NN-based graph construction can be also be viewed as a graph sparsification process which is optimizing the following objective:

$$\min_{\hat{P} \in \{0,1\}^{n \times n}} \sum_{i,j} \hat{P}_{ij} \Delta_{ij} \text{ s.t. } \sum_{j} \hat{P}_{ij} = k, \ \hat{P}_{ii} = 0, \ \forall \ 1 \leq i, j \leq n. \qquad (2.2)$$

Here, $\Delta \in \mathcal{R}_{+}^{n \times n}$ is a symmetric distance matrix. The optimal solution to the above problem is not guaranteed to be symmetric and therefore the final binary matrix P is obtained by a symmetrization step: $P_{ij} = \max(\hat{P}_{ij}, \hat{P}_{ji})$. Even though the optimization satisfies the constraints that $\sum_{j} \hat{P}_{ij} = k$ and so each node has exactly k neighbors, the subsequent symmetrization step only enforces the constraint $\sum_{j} P_{ij} \geq k$. As discussed earlier, this gives rise to irregular graphs when using k-NN based graph construction.

b-matching overcomes this problem by optimizing the following objective

$$\min_{P \in \{0,1\}^{n \times n}} \sum_{i,j} P_{ij} \Delta_{ij} \text{ s.t. } \sum_{j} P_{ij} = b, \ P_{ii} = 0, \ P_{ij} = P_{ji}, \ \forall 1 \leq i, j \leq n. \qquad (2.3)$$

It can be seen that the degree ($\sum_{j} P_{ij} = b$) and symmetrization ($P_{ij} = P_{ji}$) constraints are part of the optimization problem and thus there is no need of any ad-hoc post-processing symmetrization step as is the case in k-NN. However, the above objective is significantly harder to optimize compared to the vanilla k-NN objective presented previously. More details will be forthcoming in Section 2.2.2.

Edge Re-Weighting
Once a set of edges have been selected through the matrix P in the previous section, the goal here is to estimate the weights of the selected edges. In other words, estimate the symmetric edge weight matrix W. Jebara et al. [2009] present three different edge re-weighting strategies.

- **Binary (BN)**: Set $W = P$. Thus $W_{ij} = P_{ij}$. One possible downside of this strategy is that it prevents recovery from errors committed during estimation of P.

- **Gaussian Kernel (GK)**: Here,

$$W_{ij} = P_{ij} \exp\left(-\frac{d(\mathbf{x}_i, \mathbf{x}_j)}{2\sigma^2}\right),$$

  where $d(\mathbf{x}_i, \mathbf{x}_j)$ measures the distance between points $\mathbf{x}_i$ and $\mathbf{x}_j$, and σ is a hyper-parameter. Possible choices for $d(\cdot, \cdot)$ include: l_p distance [Zhu, 2005], χ^2 distance, and cosine distance [Belkin et al., 2005].

- **Locally Linear Reconstruction (LLR)**: This is based on the Locally Linear Embedding (LLE) technique of Roweis and Saul [2000]. The goal is to reconstruct $\mathbf{x}_i$ from its neighborhood, i.e., we would like to minimize, $||\mathbf{x}_i - \sum_{j} P_{ij} W_{ij} \mathbf{x}_j||^2$. This leads to the following optimization problem:

$$\min_{W} \sum_{i} ||\mathbf{x}_i - \sum_{j} P_{ij} W_{ij} \mathbf{x}_j||^2 \text{ s.t. } \sum_{j} W_{ij} = 1, \ W_{ij} \geq 0, \ i = 1, \dots, n. \qquad (2.4)$$

Results

Empirical evidence of importance of graph construction and its resulting effect on SSL is presented in Jebara et al. [2009]. The results show that in most cases, a graph constructed using b-matching improves over the performance of a k-NN graph. The results also show that the choice of the inference algorithm has an impact on the results thus suggesting a coupling between the process used for constructing the graph and the inference algorithm.

b-matching Complexity

While the greedy algorithm for k-NN is well known, complexity of the known polynomial time algorithm for maximum weight b-matching is $O(bn^3)$. Although recent advances in terms of loopy belief propagation has resulted in faster variants [Huang and Jebara, 2007].

Discussion

In Jebara et al. [2009], the sparse matrix P is selected with one similarity measure while the edge re-weights are estimated with respect to another similarity measure (e.g., GK and LLR) resulting in a mismatch which may lead to sub-optimal behavior. It is worthwhile to consider learning both jointly using one similarity measure, and with some added penalty in favor of sparseness.

In Jebara et al. [2009], the hyper-parameters are heuristically set which may favor one method or the other. In order to have a thorough comparison, a grid search over these parameters results based on those would have been more insightful. Also, Computational complexity is one of the major concerns for b-matching. It would have been insightful to have runtime comparison of b-matching with k-NN.

As suggested in Jebara et al. [2009], it will be interesting to see theoretical justification for the advantages of b-matching, which is currently a topic of future research.

2.2.3   GRAPH CONSTRUCTION USING LOCAL RECONSTRUCTION

Given $\{x_1, \ldots, x_n\} \in \mathcal{R}^{d \times 1}$, the goal of Daitch et al. [2009] is to fit the "right" graph for this data. They propose algorithms to learn *hard* and *α-soft* graphs and study their properties. We present salient aspects of their work.

Hard Graphs

Daitch et al. [2009] define *hard graphs* as graphs where each node is strictly required to have a weighted degree of at least 1. The weighted degree of a node i, D_i, is given by $D_i = \sum_{j \neq i} W_{ij}$. Daitch et al. [2009] propose minimizing the following objective:

$$\mathbf{C}^{(HG)}(W) = \sum_i ||D_i \mathbf{x}_i - \sum_j W_{ij} \mathbf{x}_j||^2, \text{ s.t., } D_i \geq 1, W_{ij} = W_{ji} \geq 0, \forall 1 \leq i, j \leq n \quad (2.5)$$

for learning hard graphs. Note that without the degree constraint ($D_i \geq 1$), the above problem has a trivial solution where $W_{ij} = 0, \ \forall i, j$. When $D_i = 1$, the above objective is similar to

LLE [Roweis and Saul, 2000]. While in the case of LLE, the neighbors are selected *a priori*, and the edge weight matrix may be asymmetric and have negative entries, here, neighborhood selection and edge weight estimation are done jointly. Further the weight matrix is symmetric and each entry is guaranteed to be non-negative. Daitch et al. [2009] show how the above objective may be reformulated in terms of the Laplacian of the constructed graph. However, optimizing the above objective may be computationally infeasible. Hence, an incremental (greedy) algorithm is presented in Daitch et al. [2009]. The idea is solve the quadratic program 2.5 initially for a small subset of edges and then incrementally add new edges to the problem such that they improve the hard graph, and in the process remove any edge whose weight has been set to zero. This process is repeated until the graph cannot be improved any further. New edges to be added to the program are determined based on violation of KKT conditions [Nocedal and Wright, 1999] of the Lagrangian.

α-Soft Graphs

In some cases, it may be necessary to relax the degree constraint for some nodes. For example, one may prefer sparser connectivity for outlier nodes compared to nodes in high density regions. The degree constraint in (2.5) can be relaxed to the following:

$$\sum_i (\max(0, 1 - D_i))^2 \leq \alpha n, \tag{2.6}$$

where α is a hyper-parameter. Graphs estimated with this new constraint are called α-*soft graphs*. If all nodes satisfied the hard degree requirement ($D_i \geq 1$) and thus $1 - D_i \leq 0$, $\forall i$ then the above constraint (2.6) is trivially satisfied. Each *hinge loss* term, $\max(0, 1 - D_i)$ incurs a penalty when the corresponding node $\mathbf{x}_i$ violates the degree constraint, i.e., if $D_i < 1$. Note that the value of $\mathbf{C}(W) = \sum_i ||d_i \mathbf{x}_i - \sum_j W_{ij} \mathbf{x}_j||^2$ is decreased (improved) by uniformly scaling down weights of all the edges. In that case, it is easy to verify that the constraint (2.6) will be satisfied with equality. If we define $\eta(W) = \sum_i (\max(0, 1 - D_i))^2$, then

$$\eta(W^*) = \alpha n, \tag{2.7}$$

where W^* is a minimizer of $\mathbf{C}^{(HG)}(W)$ subject to constraint (2.6). Instead of incorporating constraint (2.6) directly, the algorithm for α-soft graphs in Daitch et al. [2009] incorporates the constraint as a regularization term.

Properties of the induced Graphs

Theorem 2.3 [Daitch et al., 2009].   *For every $\alpha > 0$, every set of n vectors in $\mathcal{R}^{d \times 1}$ has a hard and an α-soft graph with at most $(d + 1)n$ edges.*

This theorem suggests that the average degree of any node in the graph is at most $2(d + 1)$. For situations where $n \gg d$, this theorem guarantees a sparse graph. Daitch et al. [2009] validate this empirically. However, this is not a very tight bound for high dimensional data, e.g., problems in Natural Language Processing where instances often have millions of dimensions (or features).

Daitch et al. [2009] also suggest (without theoretical or empirical validation) that the average degree of a hard or α-soft graph of a set of vectors is a measure of the effective dimensionality of those vectors, and that it will be small if they lie close to a low-dimensional manifold of low curvature. If it were indeed the case, then such graph construction process may be independently useful in discovering dimensionality of a set of points in high dimensional space.

Results

Daitch et al. [2009] present a variety of experimental results which demonstrate properties of the constructed hard and α-soft graphs (with $\alpha = 0.1$) as well as their effectiveness in classification, regressions and clustering tasks. The results show that, in practice, the average degree of a node in the hard and soft graphs are lower than the upper bound predicted by Theorem 2.3. It can also be observed that the computational time for graph construction increases with n. This raises scalability concerns for such methods. When used with the label inference algorithm of Zhu et al. [2003] for the graph-based SSL, both hard and 0.1-soft graphs outperform other graph construction methods.

Relationship to Linear Neighborhood Propagation (LNP) [Wang and Zhang, 2008]

The Linear Neighborhood Propagation (LNP) algorithm [Wang and Zhang, 2008] is a two-step process whose step 1 involves a graph construction step. In this case, the graph is constructed by minimizing an objective (2.8) which is similar to the LLE objective [Roweis and Saul, 2000]. As in Section 2.2.3, the difference with LLE is that in LNP, the edge weight matrix W is symmetric and that all edge weights are non-negative (these constraints are not shown in (2.8)).

$$\min_{W} \sum_i ||\mathbf{x}_i - \sum_j W_{ij}\mathbf{x}_j||^2, \text{ s.t. } d_i = \sum_j W_{ij} = 1, i = 1,\ldots,n. \tag{2.8}$$

By comparing (2.8) with (2.5) from Section 2.2.3, we note that the hard graph optimization enforces the constraint $d_i \geq 1$ while the LNP graph construction optimization (2.8) enforces $d_i = 1$. In other words, LNP is searching a smaller space of graphs compared to hard graphs.

2.3   TASK-DEPENDENT GRAPH CONSTRUCTION

2.3.1   INFERENCE-DRIVEN METRIC LEARNING (IDML)

IDML [Dhillon et al., 2010] is a semi-supervised metric learning algorithm for graph construction. We first present a review of the metric learning literature followed by the details of the IDML algorithm.

METRIC LEARNING REVIEW

Given $\mathbf{x}_i, \mathbf{x}_j \in \mathcal{R}^{d \times 1}$, the Mahalanobis distance between them is given by

$$d_M(\mathbf{x}_i, \mathbf{x}_j) = (\mathbf{x}_i - \mathbf{x}_j)^\top M(\mathbf{x}_i - \mathbf{x}_j),$$

where $M \in \mathcal{R}^{d \times d}$ is a positive semi-definite matrix and so $\mathbf{x}^\top M \mathbf{x} \geq 0 \ \forall \mathbf{x} \in \mathcal{R}^{d \times 1}$. As a result, there exists a $P \in \mathcal{R}^{d \times d}$ such that $M = P^\top P$. Thus, we have that

$$\begin{aligned}
d_M(\mathbf{x}_i, \mathbf{x}_j) &= (\mathbf{x}_i - \mathbf{x}_j)^\top M(\mathbf{x}_i - \mathbf{x}_j) \\
&= (\mathbf{x}_i - \mathbf{x}_j)^\top P^\top P(\mathbf{x}_i - \mathbf{x}_j) \\
&= (P(\mathbf{x}_i - \mathbf{x}_j))^\top (P\mathbf{x}_i - P\mathbf{x}_j) \\
&= (P\mathbf{x}_i - P\mathbf{x}_j)^\top (P\mathbf{x}_i - P\mathbf{x}_j) = \| P\mathbf{x}_i - P\mathbf{x}_j \|_2^2 .
\end{aligned} \tag{2.9}$$

As a result, computing the Mahalanobis distance w.r.t. M is equivalent to first projecting the instances into a new space using an appropriate transformation matrix P and then computing squared Euclidean distance in the linearly transformed space.

Information-Theoretic Metric Learning (ITML)

Information-Theoretic Metric Learning (ITML) [Davis et al., 2007] takes as input a inter-instance distance matrix $\Delta \in \mathcal{R}^{n \times n}$ where $[\Delta]_{ij} = d(\mathbf{x}_i, \mathbf{x}_j)$ where d is some distance metric. Instances i and j are similar if $d(\mathbf{x}_i, \mathbf{x}_j) \leq u$ and considered dissimilar if $d(\mathbf{x}_i, \mathbf{x}_j) \geq l$. Let the set the similar instance be represented by $\mathscr{S}$ while the set of dissimilar instances are represented by $\mathscr{D}$. Here, u and l are pre-determined distance thresholds.

In addition to prior knowledge about inter-instance distances, sometimes prior information about the matrix M, denoted by M_0, may also be available. In such cases, we prefer a learned matrix M to be as close as possible to the prior matrix M_0. ITML combines these two types of prior information, i.e., knowledge about inter-instance distances, and prior matrix M_0, in order to learn the matrix M by solving the following optimization problem:

$$\begin{aligned}
\min_{M \succeq 0} \quad & d_{\mathrm{ld}}(M, M_0) \\
\text{s.t.,} \quad & \mathrm{tr}\{M(\mathbf{x}_i - \mathbf{x}_j)(\mathbf{x}_i - \mathbf{x}_j)^\top\} \leq u, \ \forall (i, j) \in \mathscr{S} \\
& \mathrm{tr}\{M(\mathbf{x}_i - \mathbf{x}_j)(\mathbf{x}_i - \mathbf{x}_j)^\top\} \geq l, \ \forall (i, j) \in \mathscr{D},
\end{aligned} \tag{2.10}$$

where $d_{\mathrm{ld}}(M, M_0) = \mathrm{tr}(M M_0^{-1}) - \log \det(M M_0^{-1}) - n$ is the LogDet divergence and $M \succeq 0$ implies that M is constrained to be positive semi-definite. It may not always be possible to *exactly* solve the above optimization problem. To handle such situations, slack variables are introduced to the ITML objective. Let $c(i, j)$ be the index of the constraint between instances i and j, and let $\boldsymbol{\xi}$ be a vector of length $|\mathscr{S}| + |\mathscr{D}|$ of slack variables. $\boldsymbol{\xi}$ is initialized to $\boldsymbol{\xi}_0$, whose components equal u for similarity constraints, and l for dissimilarity constraints. The modified ITML

Algorithm 1: Inference-Driven Metric Learning (IDML) **Input**: instances X, training labels Y, training instance indicator S, label entropy threshold β, neighborhood size k
Output: Mahalanobis distance parameter M

1: $\hat{Y} \leftarrow Y, \hat{S} \leftarrow S$
2: **repeat**
3:    $M \leftarrow \text{MetricLearner}(X, \hat{S}, \hat{Y})$
4:    $W \leftarrow \text{ConstructKnnGraph}(X, M, k)$
5:    $\hat{Y}' \leftarrow \text{GraphLabelInference}(W, \hat{S}, \hat{Y})$
6:    $U \leftarrow \text{SelectLowEntInstances}(\hat{Y}', \hat{S}, \beta)$
7:    $\hat{Y} \leftarrow \hat{Y} + U\hat{Y}'$
8:    $\hat{S} \leftarrow \hat{S} + U$
9: **until** convergence (i.e., $U_{ii} = 0, \ \forall i$)
10: return M

---

objective involving slack variables is shown in (2.11).

$$\min_{M \succeq 0, \xi} \quad d_{\text{ld}}(M, M_0) + \gamma d_{\text{ld}}(\xi, \xi_0) \tag{2.11}$$
$$\text{s.t.,} \quad \text{tr}\{M(\mathbf{x}_i - \mathbf{x}_j)(\mathbf{x}_i - \mathbf{x}_j)^\top\} \leq \boldsymbol{\xi}_{c(i,j)}, \ \forall (i,j) \in \mathscr{S}$$
$$\text{tr}\{M(\mathbf{x}_i - \mathbf{x}_j)(\mathbf{x}_i - \mathbf{x}_j)^\top\} \geq \boldsymbol{\xi}_{c(i,j)}, \ \forall (i,j) \in \mathscr{D}$$

where γ is a hyper-parameter which determines the importance of violated constraints. The above optimization problem may be solved by making use of repeated Bregman projections [Davis et al., 2007].

INFERENCE-DRIVEN METRIC LEARNING (IDML)

In this section, we present Inference-Driven Metric Learning (IDML) (Algorithm 1) [Dhillon et al., 2010], a metric learning framework which uses *both* labeled and unlabeled data. It is based on a supervised metric learning algorithm, e.g., ITML. In self-training styled iterations, IDML alternates between metric learning and label inference, with output of label inference used during next round of metric learning, and so on.

IDML is based on the assumption that supervised metric learning algorithms can learn a better metric if the number of available labeled instances is increased. As the focus is on a SSL setting with n_l labeled and n_u unlabeled instances, the idea is to automatically label the unlabeled instances using a graph-based SSL algorithm, and then include instances with low assigned label entropy (i.e., high confidence label assignments) as "labeled" in the next round of metric learning. The number of instances added in each iteration depends on a hyper-parameter β. This process is continued until no new instances can be added to the set of labeled instances, which can happen when either all the instances are exhausted, or when none of the remaining unlabeled instances can be assigned labels with high confidence.

The IDML framework is presented in Algorithm 1. In step 3, any supervised metric learner, such as ITML, may be used as the MetricLearner. Using the distance metric learned in step 3, a new k-NN graph is constructed in step 4, whose edge weight matrix is stored in W. In step 5, GraphLabelInference optimizes over the graph constructed in the previous step using the Gaussian Random field (GRF) objective [Zhu et al., 2003]:

$$\min_{\hat{Y}} \quad \text{tr}\{\hat{Y}^\top L \hat{Y}\} \text{ s.t. } SY = S\hat{Y}, \tag{2.12}$$

where $S \in \mathcal{R}^{n \times n}$ is the labeled instance indicator with $S_{ii} = 1 \ \forall 1 \leq i \leq n_l$ and $S_{ij} = 0$ for all other values of $1 \leq i, j \leq n$; $Y \in \mathcal{R}_+^{n \times |\mathcal{Y}|}$ is the label matrix with the labels for the current set of labeled nodes. Recall that $L = D - W$ is the unnormalized Laplacian, and D is the diagonal matrix with $D_{ii} = \sum_j W_{ij}$. We will be presenting the above objective in more detail in Chapter 3, but briefly, the above objective requires that the labels of two nodes connected by a highly weighted edge be similar while also ensuring that the inferred labels for the "labeled" nodes do not change (as a result of the constraint $SY = S\hat{Y}$). In step 6, an unlabeled instance x_i, $1 \leq i \leq n_u$ is considered a new labeled training instance for the next round of metric learning if the instance has been assigned labels with high confidence in the current iteration, i.e., if its label distribution has low entropy (i.e., Entropy$(\hat{Y}) \leq \beta$). Finally in step 7, training instance label information is updated. This iterative process is continued till no new labeled instance can be added. IDML returns the learned matrix M which can be used to compute Mahalanobis distance using Equation 2.9. Dhillon et al. [2010] show that Algorithm 1 terminates in at most n iterations.

IDML OBJECTIVE

Intuitively, IDML optimizes a *graph-regularized metric learning* objective of the form

$$\min_{M, \hat{Y}} \quad \text{MetricLoss}(M, \hat{Y}) + \alpha \text{GraphReg}(G, W, \hat{Y}). \tag{2.13}$$

Here, MetricLoss takes the Mahalanobis distance parameter M and the current labeled information $\hat{Y}$ as inputs and measures the metric learning related loss while the second term imposes a graph-based regularization penalty. Note that the edge weight matrix W is dependent on the metric learning parameter M. α is a hyper-parameter which determines the balance between the two components of the objective. IDML can be seen as solving the above objective in an alternating minimization fashion—in one step MetricLoss is minimized keeping $\hat{Y}$ fixed, and in the second step $\hat{Y}$ is estimated keeping M and thereby W fixed.

RELATIONSHIP TO OTHER METHODS

IDML is similar in spirit to the EM-based HMRF-KMeans algorithm in Bilenko et al. [2004], which focuses on integrating constraints and metric learning for semi-supervised clustering.

Table 2.1: Comparison of transductive classification performance over graphs constructed using different methods, with $n_l = 100$ and $n_u = 1400$. All results are averaged over four trials

Datasets	Original	RP	PCA	ITML	LMNN	IDML-LM	IDML-IT
Amazon	0.4046	0.3964	0.1554	0.1418	0.2405	0.2004	**0.1265**
Newsgroups	0.3407	0.3871	0.3098	**0.1664**	0.2172	0.2136	**0.1664**
Reuters	0.2928	0.3529	0.2236	0.1088	0.3093	0.2731	**0.0999**
EnronA	0.3246	0.3493	0.2691	0.2307	0.1852	**0.1707**	0.2179
Text	0.4523	0.4920	0.4820	0.3072	0.3125	0.3125	**0.2893**
USPS	**0.0639**	0.0829	–	0.1096	0.1336	0.1225	0.0834
BCI	0.4508	0.4692	–	0.4217	0.3058	**0.2967**	0.4081
Digit	**0.0218**	0.0250	–	0.0281	0.1186	0.0877	0.0281

However, there are crucial differences. First, the inference algorithm in Bilenko et al. [2004] is parametric (k-Means), while the inference in IDML is nonparametric (graph-based). More importantly, the label constraints in Bilenko et al. [2004] are hard constraints which are fixed *a priori* and are not changed during learning. In case of IDML , the graph structure induced in each iteration (step 4 in Algorithm 1) imposes a Manifold Regularization (MR) [Belkin et al., 2005] style smoothing penalty, as shown in Equation 2.12. Hence, compared to the hard and fixed constraints in HMRF-KMeans, IDML constraints are soft, and new constraints are added in each iteration of the algorithm.

IDML is different from previous work on supervised metric learning [Davis et al., 2007, Jin et al., 2009, Weinberger and Saul, 2009] is the use of unlabeled data. Moreover, in all of the previously proposed algorithms, the constraints used during metric learning are fixed *a priori*, as they are usually derived from labeled instances which do not change during the course of the algorithm; while the constraints in IDML are adaptive and new constraints are added in each iteration, when automatically labeled instances are included in each iteration. In the next section, we present empirical evidence that these additional constraints can be quite effective in improving IDML's performance compared to ITML.

RESULTS

In this section, we present results comparing the following methods for estimating M, which in turn is used to estimate edge weights:

Original: We set $M = I_{d \times d}$, i.e., the data is not transformed and Euclidean distance in the input space is used to compute the distance between instances.

RP: The data is first projected into a lower dimensional space of dimension $d' = \frac{\log n}{\epsilon^2 \log \frac{1}{\epsilon}}$ using the Random Projection (RP) method [Bingham and Mannila, 2001]. We set $M = R^\top R$, where R is the projection matrix used by RP. ϵ is set to 0.25.

PCA: Instances are first projected into a lower dimensional space using Principal Components Analysis (PCA). For all experiments, dimensionality of the projected space was set at 250[1]. We set $M = P^\top P$, where P is the projection matrix generated by PCA.

LMNN: M is learned by applying LMNN [Blitzer et al., 2005] on the PCA projected space (above).

ITML: M is learned by applying ITML (see Section 2.3.1) on the PCA projected space (above).

IDML-LM: M is learned by applying IDML (Algorithm 1) on the PCA projected space (above); with LMNN used as METRICLEARNER in IDML.

IDML-IT: M is learned by applying IDML (Algorithm 1) (see Section 2.3.1) on the PCA projected space (above); with ITML used as METRICLEARNER in IDML.

Experimental results with 100 labeled instances (n_l) are shown in Tables 2.1. From this we observe that constructing a graph using a learned metric can significantly improve performance. We consistently find graphs constructed using IDML-IT to be the most effective. This is particularly true in case of high-dimensional datasets where distances in the original input space are often unreliable because of curse of dimensionality. Dhillon et al. [2010] report that they were unable to include comparisons on graphs constructed using b-matching [Jebara et al., 2009] as it often resulted in disconnected graphs which the GRF code they used (obtained from Zhu et al. [2003]) was unable to handle.

2.3.2 GRAPH KERNELS BY SPECTRAL TRANSFORM

The goal of graph-based SSL it to balance between *accuracy* of the labeled nodes and *smoothness* with respect to the variation of labels over the graph. A solution is said to be smooth if the labels across a highly weighted edge are similar to each other. A central quantity in determining smoothness is the graph Laplacian. Zhu et al. [2007] noted that the eigenvectors of the graph Laplacian play a critical role in determining the smoothness of a solution. Eigenvectors with small eigenvalues are smooth while those with large eigenvalues are rugged. Zhu et al. [2007] show that different weightings of eigenvectors of the graph Laplacian lead to different measures of smoothness. Such weightings are called *spectral transforms*. In this section we show effective spectral transforms maybe obtained automatically. We start with a review of the basic properties of a graph Laplacian and its relationship to smoothness.

Spectrum of the Graph Laplacian
Recall that the unnormalized Laplacian, L, of a graph $G = (V, E, W)$ is given by $L = D - W$. Here W is the symmetric edge weight matrix, D is a diagonal matrix with $D_{ii} = \sum_j W_{ij}$ and

---

[1]PCA was not performed on USPS, BCI and Digit as they already had dimension lower than 250.

$D_{ij} = 0$, $\forall i \neq j$. $L, D,$. Given a function, $f : V \to R$, which assigns scores (for a particular label) to each vertex in V, it is easy to verify that

$$\mathbf{f}^T L \mathbf{f} = \sum_{i,j \in V} W_{ij}(f(i) - f(j))^2, \tag{2.14}$$

where $\mathbf{f} = [f(1), \dots, f(|V|)]^\top$. We note that minimization of Equation 2.14 enforces the smoothness condition as it requires that $f(i) \approx f(j)$ when W_{ij} is large.

As $W_{ij} \geq 0$, $\forall i, j$ we have that $\mathbf{f}^T L \mathbf{f} \geq 0$, $\forall \mathbf{f}$ which implies that L is a positive semi-definite matrix. Let $\lambda_1 \leq \lambda_2 \leq \dots \leq \lambda_n$ and $\phi_1, \phi_2, \dots \phi_n \in \mathcal{R}^{n \times 1}$ be the set of eigenvalues and orthonormal eigenvectors of L, respectively. The spectral decomposition of the Laplacian is given by $L = \sum_i \lambda_i \phi_i \phi_i^\top$.

Consider

$$L\phi_i = \lambda_i \phi_i \text{ and so}$$
$$\phi_i^\top L \phi_i = \lambda_i \phi_i^\top \phi_i = \lambda_i. \tag{2.15}$$

As a result, if $\mathbf{f} = \phi_i$, then the value of $\mathbf{f}^T L \mathbf{f}$ would be λ_i. In order words, the smoothness penalty incurred by a particular eigenvector, ϕ_i, is then given by λ_i. As a result, eigenvectors with smaller eigenvalues are smoother. For illustration, consider a simple graph (with two connected components) and its spectral decomposition shown in Figure 2.3. We observe that:

- the first two eigenvalues λ_1 and λ_2 are 0. This is equal to the number of connected components in the graph, i.e., 2. In general, a graph has k connected components iff $\lambda_i = 0 \; \forall \; 1 \leq i \leq k$. Moreover, the corresponding eigenvectors are constant within the connected components, and

- the eigenvectors are increasingly irregular with increasing eigenvalues.

This association between smoothness and ranked eigenvalues (and corresponding eigenvectors) will be exploited for learning smooth kernels over the graph.

Kernels by Spectral Transforms
In order to learn a kernel matrix $K \in \mathbb{R}^{n \times n}$ which penalizes functions which are not smooth over the graph, Zhu et al. [2007] propose the following kernel form:

$$K = \sum_i \mu_i \phi_i \phi_i^T, \tag{2.16}$$

where ϕ_i's are the eigenvectors of the graph Laplacian L, and $\mu_i \geq 0$ are the eigenvalues of K. Note that K is a non-negative sum of outer products and hence a positive semi-definite matrix and thus a valid kernel matrix. It is important to point out that in order to estimate K, an initial graph structure is necessary as an estimation of K is dependent on the graph Laplacian.

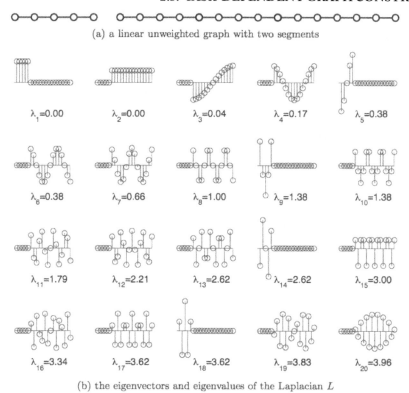

(a) a linear unweighted graph with two segments

$\lambda_1=0.00$ $\lambda_2=0.00$ $\lambda_3=0.04$ $\lambda_4=0.17$ $\lambda_5=0.38$

$\lambda_6=0.38$ $\lambda_7=0.66$ $\lambda_8=1.00$ $\lambda_9=1.38$ $\lambda_{10}=1.38$

$\lambda_{11}=1.79$ $\lambda_{12}=2.21$ $\lambda_{13}=2.62$ $\lambda_{14}=2.62$ $\lambda_{15}=3.00$

$\lambda_{16}=3.34$ $\lambda_{17}=3.62$ $\lambda_{18}=3.62$ $\lambda_{19}=3.83$ $\lambda_{20}=3.96$

(b) the eigenvectors and eigenvalues of the Laplacian L

Figure 2.3: A linear-chain graph (part (a)) with two segments and its spectral decomposition (part (b)). Note that the eigenvectors are smoother for lower eigenvalues [Zhu et al., 2007].

To assign a higher weight, μ_i, to the outer product $\phi_i \phi_i^T$ of a smoother eigenvector ϕ_i (and correspondingly smaller eigenvalue λ_i) of L, a *spectral transformation* function $r : \mathbb{R}_+ \rightarrow \mathbb{R}_+$ is defined, where r is non-negative and decreasing. By setting $\mu_i = r(\lambda_i)$, we have that

$$K = \sum_i r(\lambda_i)\phi_i \phi_i^T. \tag{2.17}$$

The spectral transformation essentially reverses the order of the eigenvalues, so that smooth eigenvectors have a larger contribution to K.

There is a question, however, about the choice of the function r. Chapelle et al. [2002] and Smola and Kondor [2003] suggest a number of possibilities include setting $r(\lambda_i) = \frac{1}{\lambda_i + \epsilon}$ (where ϵ is a hyper-parameter). This leads to the popular Gaussian field kernel used in Zhu et al. [2003]. However, while there are many natural choices for r, it may not be clear up front which is most relevant for a given learning task. Moreover, tuning any hyper-parameter associated with any parametric form of r is another issue. Zhu et al. [2007] address these issues by learning a spectral transformation that optimizes kernel alignment to the labeled data which we will discuss in the next section.

Optimizing Kernel Alignment

Let K_{tr} be the $l \times l$ sub-matrix of the full kernel matrix K, with K_{tr} corresponding to the l labeled instances. Let $T \in \mathcal{R}^{l \times l}$ be derived from the n_l labeled instances $\{(\mathbf{x}_i, y_i)\}_{i=1}^{n_l}$ such that $T_{ij} = 1$ if $y_i = y_j$ and -1 otherwise. Empirical kernel alignment [Cristianini et al., 2001, Lanckriet et al., 2004] between K_{tr} and T is defined as

$$A(K_{tr}, T) = \frac{\langle K_{tr}, T \rangle_F}{\sqrt{\langle K_{tr}, K_{tr} \rangle_F \langle T, T \rangle_F}}, \qquad (2.18)$$

where $< \cdot, \cdot >_F$ is the Frobenius product.[2] Note that $A(K_{tr}, T)$ is maximized when $K_{tr} \propto T$.

Zhu et al. [2007] propose an objective to estimate the kernel matrix K by maximizing the kernel alignment objective in Equation 2.18. This objective is defined directly over the transformed eigenvalues μ_i, without any parametric assumption for r which would have otherwise linked μ_i with the corresponding eigenvalue λ_i of L. Since the optimization is directly over the μ_i's, we need a way to encode the constraint that μ_i with lower index i should have higher value (Section 2.3.2). In Zhu et al. [2007], following *order constraints* $\mu_i \geq \mu_{i+1}$, $i = 1, 2, \ldots, n-1$ are added to the optimization problem to achieve this goal.

The estimated kernel matrix K should be a positive semi-definite matrix and hence the kernel alignment based optimization described above can be set up as a Semi-Definite Program (SDP). However, SDP optimization suffers from high computation complexity [Boyd and Vandenberghe, 2004]. Fortunately, the optimization problem can also be set up in a more computationally efficient form. The resulting maximization problem is

$$\mathbf{C}^{(OC)} = A(K_{tr}, T) \text{ s.t. } K = \sum_i \mu_i \phi_i \phi_i^T, \text{Tr}(K) = 1$$

$$\mu_i \geq 0, \ \forall 1 \leq i \leq n; \mu_i \geq \mu_{i+1}, \ \forall 1 \leq i \leq n-1.$$

Note that as $\mu_i \geq 0$ and $\phi_i \phi_i^T$ are outer products, K is guaranteed to be positive semi-definite and thus a valid kernel matrix. The trace constraint is needed above to fix the scale invariance of the kernel alignment [Zhu et al., 2007]. The kernel obtained by solving the above optimization problem is called *Order Constrained Kernel* [Zhu et al., 2007], which we will denote by K_{OC}. Another property of the Laplacian is exploited in Zhu et al. [2007] to define an *Improved Order Constrained Kernel*, which we shall refer to as K_{IOC}. In particular, $\lambda_1 = 0$ for any graph Laplacian [Chung, 1997]. Moreover, if there are *no disjoint components* in the graph then ϕ_1 is constant overall[3]. In that case, $\phi_1 \phi_1^T$ is a constant matrix which does not depend on L, with the product

---

[2]$\langle M, N \rangle_F = \sum_{ij} M_{ij} N_{ij}$
[3]Otherwise, if a graph has k connected components, then the first k eigenvectors are piecewise constant over the components.

$\mu_1 \phi_1 \phi_1^T$ acting as a bias term in the definition of K. Thus, the ordering constraints above can be replaced with

$$\mu_i \geq \mu_{i+1}, \; i = 1, 2, \ldots, n - 1, \text{ and } \phi_i \text{ not constant overall} \qquad (2.19)$$

We would like to point out that constraint (2.19) is meant to impact only μ_1 and that too in connected graphs only, as in all other cases ϕ_i is not going to be constant.

Results

Results from a large number of experiments comparing K_{OC} and K_{IOC} to six other standard kernels: Gaussian [Zhu et al., 2003], Diffusion [Smola and Kondor, 2003], Max-Align [Lanckriet et al., 2004], Radial Basis Function (RBF), and Linear and Quadratic are presented in Zhu et al. [2007] . Out of these, RBF, linear and quadratic are unsupervised while the rest (including the order kernels) are semi-supervised in nature. All eight kernels are combined with an SVM for classification [Burges, 1998]. Accuracy of the SVMs on unlabeled data is used as the evaluation metric. Experimental results from 7 different datasets are reported. For each dataset, an unweighted 10-NN (with the exception of one 100-NN) graph is constructed using either Euclidean or cosine similarity. In all cases, the graphs are connected and hence constraint (2.19) is used while estimating K_{IOC}. Smallest 200 eigenvectors of Laplacians are used in all the experiments. For each dataset, 5 training set sizes are used and for each set 30 random trials are performed. From the experimental results presented in Zhu et al. [2007], we observe that the improved order constrained kernel, K_{IOC}, outperformed all other kernels.

Discussion

In the sections above, we reviewed the kernel-alignment based spectral kernel design method described in Zhu et al. [2007]. At one end, we have the *maximum-alignment* kernels [Lanckriet et al., 2004] and at the other end we have the parametric Gaussian field kernels [Zhu et al., 2003]. The order-constrained kernels introduced in Zhu et al. [2007] strikes a balance between these two extremes.

Relationship to [Johnson and Zhang, 2008]

In Zhu et al. [2007] and its review in the sections above, we have seen how to derive a kernel from the graph Laplacian and effectiveness of the derived kernel for supervised learning. Johnson and Zhang [2008] look at the related issue of the relationship between graph-based semi-supervised learning and supervised learning. Let k be a given a kernel function and K the corresponding kernel matrix. Let $L = K^{-1}$ be the Laplacian of a graph. In that case, in Theorem 3.1 of Johnson and Zhang [2008], it is proved that graph-based semi-supervised learning using Laplacian $L = K^{-1}$ is the equivalent to a kernel-based supervised learner (e.g., SVM) with kernel function k.

In Zhu et al. [2007], spectrum of the Laplacian of a fixed graph is modified to compute a kernel. While in Johnson and Zhang [2008], the kernel is created up front using all the data and without need for any a-priori fixed graph structure.

2.4   CONCLUSION

In this chapter, we reviewed several recently proposed methods for graph-construction [Daitch et al., 2009, Jebara et al., 2009, Wang and Zhang, 2008]. Locally Linear Reconstruction [Roweis and Saul, 2000] based edge weighting and enforcement of degree constraint on each node is a common theme in these methods. Daitch et al. [2009] further analyze properties of the induced graphs, while Jebara et al. [2009] emphasize the need for regular graphs in graph-based SSL. Additionally, we also looked at semi-supervised spectral kernel design [Zhu et al., 2007]. Even though the graph structure is fixed a-priori in this case, this method highlights useful properties of the graph Laplacian and also demonstrates the relationship between the graph structure and a kernel that can be derived from it. A few steps for future work on graph construction for graph-based SSL is discussed in Chapter 6.

CHAPTER 3

Learning and Inference

Once the graph is constructed, the next step in solving an SSL problem using graph-based methods is the injection of seed labels on a subset of the nodes in the graph followed by the process of inferring the labels for the unlabeled nodes. In this chapter, we first examine the design choices involved in this seed labeling process. We then present a number of approaches for label inference. While a majority of the graph-based inference approaches are transductive, there are a some inductive graph-based SSL approaches as well.

3.1 SEED SUPERVISION

The seed label information is specified using a matrix $Y \in \mathcal{R}_+^{n \times m}$, where Y_{vl} specifies the value of seed label l for node v. Note that this representation of seed labels allows for both multi-class and multi-label seeds. While some methods require seed label scores to be a probability distribution [Subramanya and Bilmes, 2010], i.e., each row of Y should be a probability distribution, others do not have such a requirement [Orbach and Crammer, 2012, Talukdar and Crammer, 2009]. Sometimes the seed labels may themselves be noisy; in order to handle these cases, some algorithms allow the seed labels to change during inference while others, such as Blum and Chawla [2001] and Zhu et al. [2003], hold them fixed.

3.2 TRANSDUCTIVE METHODS

As explained previously, given a training set consisting of labeled and unlabeled data, the goal of a transductive algorithm is to learn a function that is able to predict the labels for *only* the unlabeled data. A majority of the graph-based SSL algorithms are transductive.

3.2.1 GRAPH CUT

Blum and Chawla [2001] proposed one of the first graph-based SSL algorithms. They posed SSL as a graph mincut (sometimes also referred to as *st-cut*) problem. Their approach applies to only binary classification problems with the positive labeled vertices acting as sources while the negatively labeled vertices are sinks. It can be shown that the objective they minimize is given by:

$$\operatorname*{argmin}_{\hat{Y} \in \{+1,-1\}^n} \sum_{(u,v) \in E} W_{uv} |\hat{Y}_u - \hat{Y}_v|$$
$$\text{s.t. } Y_u = \hat{Y}_u, \ \forall u \text{ where } S_{uu} = 1$$

with $S \in \mathcal{R}_+^{n \times n}$ being the seed label indicator matrix with $S_{ii} = 1$ if $1 \leq i \leq n_l$ and 0 otherwise. Instead of storing label scores in matrix $\hat{Y}$ as defined before, $\hat{Y}$ in this case is a vector of length n with the label identity stored in each entry. In other words, $\hat{Y} \in \{+1, -1\}^n$ and thus each $\hat{Y}_u \in \{+1, -1\}$. Further, $|\mathcal{Y}| = 2$, and the min-cut objective is a discrete optimization problem.

The above objective finds a minimum set of edges whose removal blocks all flow from the sources to the sinks and thus the name *mincut*. Blum and Chawla [2001] show that the mincut is the mode of a Markov random field with binary labels. One of the problems with mincut is that it only provides hard classification without confidence. To address this, Blum et al. [2004] propose a variant where they perturb the graph by adding random noise to the edge weights and apply mincut to the multiple (perturbed) graphs with the labels being determined by a majority vote.

It has also been observed that mincut can often lead to degenerate solutions, i.e., solutions in which all the unlabeled vertices are classified as belonging to a single class (either positive or negative). To address this, Joachims [2003] modify the mincut with a normalized objective

$$\underset{\hat{Y} \in \{+1,-1\}^n}{\arg\min} \frac{\sum_{(u,v) \in E} W_{uv} |\hat{Y}_u - \hat{Y}_v|}{|u : \hat{Y}_u = +1||u : \hat{Y}_u = -1|}$$
$$\text{s.t. } Y_u = \hat{Y}_u, \ \forall u \text{ where } S_{uu} = 1 \quad .$$

This is related to the normcut problem which is known to be NP hard [Shi and Malik]. Joachims [2003] approximate it by relaxing it and optimizing soft labels rather than hard constraints. They make use of spectral methods for optimizing the above objective leading to *spectral graph transduction* (SGT). We note that such graph-cut based approaches are intrinsically meant for binary classification problems, with multi class extensions possible through one-vs.-rest setting.

3.2.2   GAUSSIAN RANDOM FIELDS (GRF)

Gaussian Random Fields (GRF) is another example of the early work in the area of graph-based SSL [Zhu et al., 2003]. GRF solves the following optimization problem:

$$\underset{\hat{Y} \in \mathcal{R}^n}{\arg\min} \sum_{l \in \mathcal{Y}} \sum_{(u,v) \in E} W_{uv} (\hat{Y}_{ul} - \hat{Y}_{vl})^2$$
$$\text{s.t. } Y_{ul} = \hat{Y}_{ul}, \ \forall l \in \mathcal{Y}, \ \forall u \text{ for which } S_{uu} = 1$$

where $S \in \mathcal{R}_+^{n \times n}$ is the seed label indicator matrix with $S_{ii} = 1$ if $1 \leq i \leq n_l$ and 0 otherwise. Thus, GRF aims at obtaining a smooth label assignment over the graph, while keeping the labels on the seed nodes unchanged. The above optimization problem can also be alternatively written in matrix form as follows:

$$\underset{\hat{Y} \in \mathcal{R}^n}{\arg\min} \sum_{l \in \mathcal{Y}} \hat{Y}_l^T L \hat{Y}_l$$
$$\text{s.t. } SY = S\hat{Y}.$$

The GRF objective can be efficiently optimized using the following iterative update:

$$\hat{Y}_{vl}^{(t+1)} \leftarrow \frac{\sum_{(u,v)\in E} W_{uv} \hat{Y}_{ul}^{(t)}}{\sum_{(u,v)\in E} W_{uv}}.$$

These updates are guaranteed to converge. The resulting algorithm is more popularly known as *label propagation* (LP) as it has the notion of propagation of label information from the initially seeded (labeled) nodes to the unlabeled nodes. Please note that these updates satisfy the *harmonic property*: the score of a label assigned to a node is the weighted average of score of the label assigned to its neighbors. The use of a Gaussian kernel to construct the graph in the original GRF paper [Zhu et al., 2003] is the source of the word "Gaussian" in GRF. However, we note that GRF is more of an inference algorithm, and can be applied to arbitrary graphs, and not necessarily restricted to those constructed using Gaussian kernels alone.

3.2.3   LOCAL AND GLOBAL CONSISTENCY (LGC)

Local and Global Consistency (LGC) [Zhou et al., 2004] solves the following optimization problem:

$$\underset{\hat{Y}\in\mathcal{R}^n}{\text{argmin}} \sum_{l\in\mathcal{Y}} \left[\mu \sum_v \left(\hat{Y}_{vl} - Y_{vl} \right)^2 + \sum_{(u,v)\in E} W_{uv} \left(\frac{\hat{Y}_{ul}}{\sqrt{D_{uu}}} - \frac{\hat{Y}_{vl}}{\sqrt{D_{vv}}} \right)^2 \right].$$

Compared to the GRF objective above, LGC has two important differences: (a) the inferred labels for the "labeled" nodes are no longer required to be exactly equal to the seed values and this helps with cases where there may be noise in the seed labels, and (b) the label for each node is penalized by the degree of that node ($\hat{Y}_{ul}/\sqrt{D_{uu}}$) ensuring that in the case of irregular graphs, the influence of high degree nodes is regularized. Zhou et al. [2004] show that the above objective has the following closed form solution.

$$\hat{Y} = \frac{\mu}{1+\mu} \left(I - \frac{1}{1+\mu} D^{-1/2} W D^{-1/2} \right)^{-1} Y.$$

Zhou et al. [2004] also show that the objective may be solved using the following iterative update

$$\hat{Y}^{(t+1)} \leftarrow \frac{1}{1+\mu} \left(D^{-1/2} W D^{-1/2} \right) \hat{Y}^{(t)} + \frac{\mu}{1+\mu} Y$$

which is guaranteed to converge.

3.2.4   ADSORPTION

Adsorption [Baluja et al., 2008] is a general framework for transductive learning. Given a set of labeled and a relatively large number of unlabeled examples, the goal is to label all the unlabeled instances, and possibly, under the assumption of label-noise, also to re-label the labeled examples. There are two ways to explain the adsorption algorithm.

Random-Walk View   The Adsorption algorithm can be viewed as a controlled random walk over a graph G. The control is formalized via three possible actions denoted by *inj, cont, abnd* with pre-defined probabilities $p_v^{inj}, p_v^{cont}, p_v^{abnd} \geq 0$ for each vertex $v \in V$ such that $p_v^{inj} + p_v^{cont} + p_v^{abnd} = 1$. To label a vertex $v \in V$, we start a random walk at v with three possible transitions: with probability p_v^{inj} the random-walk stops and returns (i.e., *inject*) a pre-defined vector Y_v. We set $p_v^{inj} = 0$ for unlabeled vertices v. Second, with probability p_v^{abnd} the random-walk *abandons* the labeling process and returns the all-zeros vector $\mathbf{0}_m$. Third, with probability p_v^{cont} the random *continues* to one of v's neighbors v' with probability proportional to $W_{vv'}$. Note that by definition $W_{vv'} = 0$ if $(v, v') \notin E$. We can summarize the above process with the following transition probabilities:

$$p\left(v'|v\right) = \begin{cases} \frac{W_{v'v}}{\sum_u W_{uv}} & \text{if } (v', v) \in E \\ 0 & \text{otherwise.} \end{cases} \tag{3.1}$$

As a result the (expected) label $\hat{Y}_v$ for node $v \in V$ is given by,

$$\hat{Y}_v = \left(p_v^{inj} \times Y_v\right) + \left(p_v^{cont} \times \sum_{v'} p\left(v'|v\right) \hat{Y}_{v'}\right) + p_v^{abnd} \times \mathbf{0}_m. \tag{3.2}$$

Averaging View   We define a new class with a dummy label denoted by $\nu \notin \mathcal{Y}$. This additional label encodes ignorance about the correct label. As a result we need to add an additional row to all the vectors defined above, i.e., $Y_v, \hat{Y}_v \in \mathcal{R}_+^{m+1}$ and so $Y, \hat{Y} \in \mathcal{R}_+^{n \times (m+1)}$. We set $Y_{vv} = 0$, that is, *a priori* no vertex is associated with the dummy label, and replace the zero vector $\mathbf{0}_m$ with the vector $\mathbf{r} \in \mathcal{R}_+^{m+1}$ where $\mathbf{r}_i = 0$ for $i \neq \nu$ and $\mathbf{r}_\nu = 1$. In words, if the random-walk is abandoned, then the corresponding labeling vector is zero for all the labels in $\mathcal{Y}$, and an arbitrary value of unit for the dummy label ν. This way, there is always a non-negative score for at least one label, either a "real" one or the dummy label.

The averaging view then defines a set of *fixed-point* equations to update the predicted labels. A summary of the equations is shown in Algorithm 2. The algorithm is run until convergence. This usually means running the algorithms until the label distribution at each nodel in the graph ceases to change within some tolerance value. In practice, however, the algorithm is run for a fixed number of iterations. Baluja et al. [2008] show that modulo the dummy label, the random walk and averaging views are equivalent. As adsorption is memoryless, it easily scales to tens of millions of nodes with dense edges and can also be parallelized [Baluja et al., 2008].

Thus far we have presented two way of looking at the adsorption algorithm and presented the update equations. As it can be seen, the update equations for adsorption are relatively simple. However the algorithm has three hyperparameters namely, p_v^{inj}, p_v^{cont}, and p_v^{abnd}. Here is one possible recipe for setting their values that follows from Baluja et al. [2008] and Talukdar et al. [2008]. For each node v we define two quantities: c_v and d_v and let

$$p_v^{cont} \propto c_v \quad \text{and} \quad p_v^{inj} \propto d_v \ .$$

Algorithm 2: Adsorption Algorithm

Input:
- **Graph:** $G = (V, E, W)$
- **Prior labeling:** $Y_v \in \mathcal{R}^{m+1}$ for $v \in V$
- **Probabilities:** $p_v^{inj}, p_v^{cont}, p_v^{abnd}$ for $v \in V$

Output:
- **Label Scores:** $\hat{Y}_v$ for $v \in V$

1: $\hat{Y}_v \leftarrow Y_v$ for $v \in V$ {Initialization}
2: **repeat**
3:     $D_v \leftarrow \dfrac{\sum\limits_u W_{uv} \hat{Y}_u}{\sum\limits_u W_{uv}}$ for $v \in V$
4:     **for all** $v \in V$ **do**
5:         $\hat{Y}_v \leftarrow p_v^{inj} \times Y_v + p_v^{cont} \times D_v + p_v^{abnd} \times \mathbf{r}$
6:     **end for**
7: **until** convergence

The first quantity $c_v \in [0, 1]$ is monotonically decreasing with the number of neighbors for node v in the graph G. Intuitively, the higher the value of c_v, the lower the number of neighbors of vertex v, and hence higher the information they contain about the labeling of v. The other quantity $d_v \geq 0$ is monotonically increasing with the entropy (for labeled vertices), and in this case we prefer to use the prior-information rather than the computed quantities from the neighbors. Specifically we first compute the Shannon entropy of the transition probabilities for each node,

$$H(v) = -\sum_u p\left(u|v\right) \log p\left(u|v\right) ,$$

and then pass it through the following monotonically decreasing function

$$f(x) = \frac{\log \beta}{\log(\beta + e^x))} .$$

Note that $f(0) = log(\beta)/\log(\beta + 1)$ and that $f(x) \to 0$ as $x \to \infty$. We define

$$c_v = f\left(H(v)\right) .$$

Next,

$$d_v = \begin{cases} (1 - c_v) \times \sqrt{H(v)} & \text{if } v \text{ is labeled} \\ 0 & \text{otherwise.} \end{cases}$$

Finally, to ensure proper normalization of p_v^{cont}, p_v^{inj}, and p_v^{abnd}, we define

$$z_v = \max(c_v + d_v, 1),$$

and

$$p_v^{cont} = \frac{c_v}{z_v} \quad ; \quad p_v^{inj} = \frac{d_v}{z_v} \quad ; \quad p_v^{abnd} = 1 - p_v^{cont} - p_v^{abnd}.$$

Thus, abandonment occurs only when the continuation and injection probabilities are low enough.

Note that neighborhood entropy based estimation of the three random walk probabilities above is not necessarily the only possible to set these hyperparameters. Depending on the domain and the classification task at hand, other appropriate estimation choices may be used. However, it is noteworthy that, through these three per-node probabilities, Adsorption allows us to control the amount of information flowing through each node, and thereby facilitating greater control during label propagation.

3.2.5   MODIFIED ADSORPTION (MAD)

While there are multiple possible explanations of the Adsorption algorithm, there was a lack of clear understanding of the objective it optimizes. Knowledge of such an objective is often helpful as it allows principled extensions to the core algorithm. Talukdar and Crammer [2009] show that there is no function whose local minima would be same as the output of Adsorption. They also formulated a graph-based SSL algorithm with a well defined objective which retains most of the favourable properties of the Adsorption algorithm. The resulting algorithm is called *Modified Adsorption* (MAD) and it minimizes the following objective:

$$\mathbf{C}^{(MAD)}(\hat{Y}) = \sum_l \left[\mu_1 \left(Y_l - \hat{Y}_l \right)^\top S \left(Y_l - \hat{Y}_l \right) \; + \; \mu_2 \hat{Y}_l^T \, L \, \hat{Y}_l \; + \; \mu_3 \left\| \hat{Y}_l - R_l \right\|^2 \right]. \quad (3.3)$$

Here, $\mu_1, \mu_2, \mu_3 > 0$ are hyperparameters that determine the relative importance of each term in the above objective. Here, $\hat{Y}_l, Y_l, R_l$ are the lth columns (each of size $1 \times n$) of the matrices $\hat{Y}$, Y, and R, respectively. Details of how to solve MAD's objective is presented in Appendix B. The MAD algorithm is summarized in Algorithm 3. Please note that for a graph G which is invariant to permutations of the vertices, along with $\mu_1 = 2\mu_2 = \mu_3 = 1$, the MAD and Adsorption algorithms are exactly the same.

Discussion: Comparing Algorithm 2 (Adsorption) and Algorithm 3 (MAD) we notice a few differences. First, both algorithms are normalized in different ways: in Algorithm 2 only D_v is normalized (line 4), while in Algorithm 3 the entire value of $\hat{Y}_v$ is normalized with M_{vv} (line 6). Second, the normalization of Algorithm 3 (line 2) is larger than the normalization of Algorithm 2 (line 4), as the former equals the later plus a positive quantity. Third, the Algorithm 2 uses the possibly non-symmetric matrix W without using symmetrization (lines 4 and 6), while Algorithm 2 uses only the symmetric part of W. Since p_v^{inj} is used to effectively ignore nodes with large

---

Algorithm 3: Modified Adsorption (MAD) Algorithm

Input:
- **Graph:** $G = (V, E, W)$
- **Prior labeling:** $Y_v \in \mathcal{R}^{m+1}$ for $v \in V$
- **Probabilities:** $p_v^{inj}, p_v^{cont}, p_v^{abnd}$ for $v \in V$

Output:
- **Label Scores:** $\hat{Y}_v$ for $v \in V$

1: $\hat{Y}_v \leftarrow Y_v$ for $v \in V$ {Initialization}
2: $M_{vv} \leftarrow \mu_1 \times p_v^{inj} + \mu_2 \times p_v^{cont} \times \sum_u W_{vu} + \mu_3$
3: **repeat**
4:     $D_v \leftarrow \sum_u \left(p_v^{cont} W_{vu} + p_u^{cont} W_{uv} \right) \hat{Y}_u$
5:     **for all** $v \in V$ **do**
6:         $\hat{Y}_v \leftarrow \frac{1}{M_{vv}} \left(\mu_1 \times p_v^{inj} \times Y_v + \mu_2 \times D_v + \mu_3 \times p_v^{abnd} \times R_v \right)$
7:     **end for**
8: **until** convergence

---

degree, the non-symmetric version of Algorithm 2 ignores only large-degree neighbor of a node, while the symmetric version of Algorithm 3 ignores the neighbors of a node completely if it is of large-degree. In other words, the Adsorption algorithm does not take into account large-degree neighbors of a node, while the MAD algorithm does not let the labeling of labeled vertices with large degree to be influenced by their neighbors. Finally, note that for graphs G that are invariant to permutations of the vertices, and setting $\mu_1 = 2\mu_2 = \mu_3 = 1$ the two algorithms coincide.

Convergence: A sufficient condition for the iterative process of Equation (B.2) to converge is that M is strictly diagonally dominant [Saad, 2003], i.e.,

$$|M_{vv}| > \sum_{u \neq v} |M_{vu}| \quad \text{for all values of } v.$$

In our case here we have that

$$
\begin{aligned}
|M_{vv}| - \sum_{u \neq v} |M_{vu}| &= \mu_1 \times p_v^{inj} + \mu_2 \times \sum_{u \neq v} \left(p_v^{cont} \times W_{vu} + p_u^{cont} \times W_{uv} \right) + \mu_3 - \\
&\quad \mu_2 \times \sum_{u \neq v} \left(p_v^{cont} \times W_{vu} + p_u^{cont} \times W_{uv} \right) \\
&= \mu_1 \times p_v^{inj} + \mu_3.
\end{aligned}
\tag{3.4}
$$

Note that $p_v^{inj} \geq 0$ for all v and that μ_3 is a free parameter in (3.4). Thus, we can guarantee a strict diagonal dominance (and hence convergence) by setting $\mu_3 > 0$.

Extensions: In many machine learning settings, labels are often not mutually exclusive. For example, in settings with large number of labels (e.g., Open Domain Information Extraction), labels are likely to be dependent on one another. Talukdar and Crammer [2009] propose Modified Adsorption with Dependent Labels (MADDL), a MAD extension, which can learn with such dependent (e.g., synonymous) labels. As in MAD, MADDL's objective is also convex, which can be solved efficiently using iterative and parallelizable updates. Talukdar et al. [2012] propose GraphOrder, a novel graph-based SSL algorithm for ranking as opposed to classification which is the focus of most graph-based SSL algorithms. GraphOrder is also based on MAD, and retains its desirable properties such as convex objective and efficient and parallelizable optimization.

3.2.6   QUADRATIC CRITERIA (QC)

The objectives used in MAD and the Quadratic Criteria (QC) [Bengio et al., 2007] are related. However, unlike in the case of MAD, there is no separate *dummy label* in QC. This limits QC's ability to express ignorance of label for a given node. In addition, MAD can reduce the importance of certain nodes through the use of the three node-specific random walk probabilities. In contrast, QC does not provide a framework to discount nodes during inference. MAD can be considered a generalization of QC. In particular, as MAD may be used to simulate QC by setting $p_v^{cont} = 1$, $p_v^{inj} = p_v^{abnd} = 0$, $\forall v \in V$, and by dropping the dummy label from consideration.

3.2.7   TRANSDUCTION WITH CONFIDENCE (TACO)

The Adsorption algorithms discussed above makes it possible to reduce the importance of certain nodes during inference. One downside of the above algorithms is that the node discounting strategy has to be determined *a priori* and is not adaptive. For example, a common strategy proposed in previous applications of these algorithms is to discount nodes in proportion to their degree (or neighborhood entropy). While this has been found to be useful in many applications, it is not clear such a strategy provides the best possible performance.

Transduction with Confidence (TACO) [Orbach and Crammer, 2012] relaxes this limitation and instead relies on a data-driven strategy to discount nodes. Rather than only making use of the degree of a node, TACO takes into consideration the agreement of label assignment between a node and its neighbors. In particular, TACO reduces influence of a high-degree node only if there is high-level of disagreement among its neighbors. TACO introduces the notion of node and label-specific *confidence* to measure the level of disagreement between a node and its neighbors with regard to assignment of the label in the neighborhood.

TACO defines a per-node uncertainty matrix $\Sigma_v = \text{diag}(\sigma_{v,1}, \ldots, \sigma_{v,m}) \in \mathcal{R}_+^{m \times m}$, where $\sigma_{v,l}$ is the uncertainty regarding assignment of label l on node v. As before, $\hat{Y}_{vl}$ is the corresponding label assignment score. TACO minimizes the following objective

$$
\mathbf{C}^{TACO}(\hat{Y}, \{\Sigma_v\}) = \sum_{l \in \mathcal{Y}} \left[\frac{1}{2} \sum_v S_{vv} \left[(\frac{1}{\sigma_{vl}} + \frac{1}{\gamma}) \, (\hat{Y}_{vl} - Y_{ul})^2 \right] \right.
$$
$$
\left. + \frac{1}{4} \sum_{(u,v) \in E} W_{uv} \left[(\frac{1}{\sigma_{ul}} + \frac{1}{\sigma_{vl}}) \, (\hat{Y}_{ul} - \hat{Y}_{vl})^2 \right] + \alpha \sum_v \sigma_{vl} - \beta \sum_v \log \sigma_{vl} \right] \quad (3.5)
$$

where $\alpha, \beta, \gamma \in \mathbb{R}$ are hyperparameters. The first term encourages the estimated label scores on the seed nodes ($S_{vv} = 1$) to be close to the initial injected label scores, with a discounting factor which is inversely proportional to the uncertainty term (σ_{vl}). The second term enforces the smoothness assumption across an edge, again discounted by a factor which is inversely proportional to the uncertainties of the label assignment on the two nodes across the edge. The third and fourth terms in the objective provide regularization in terms of the uncertainties: the third term prefers the uncertainties to be finite, while the fourth term prefers non-zero uncertainties. Please note that the TACO objective above is convex in all arguments. We observe that in the case with diagonal confidence matrices as above, the TACO objective can be decomposed into m independent and smaller optimization problems, one for each label. As in MAD, TACO assignment of label scores on a node are not normalized. [Orbach and Crammer, 2012] derive efficient iterative updates to minimize (3.5):

$$
\hat{Y}_{vl}^{(t+1)} \leftarrow \frac{\sum_{(u,v) \in E} \left[W_{uv} \left(\frac{1}{\sigma_{ul}^{(t)}} + \frac{1}{\sigma_{vl}^{(t)}} \right) \hat{Y}_{vl}^{(t)} \right] + S_{v,v} \left(\frac{1}{\sigma_{vl}} + \frac{1}{\gamma} \right) Y_{vl}}{\sum_{(u,v) \in E} \left[W_{uv} \left(\frac{1}{\sigma_{ul}^{(t)}} + \frac{1}{\sigma_{vl}^{(t)}} \right) \right] + S_{v,v} \left(\frac{1}{\sigma_{vl}} + \frac{1}{\gamma} \right)} \quad (3.6)
$$

and

$$
sigma_{vl}^{(t+1)} \leftarrow \frac{\beta}{2\alpha} + \frac{1}{2\alpha} \sqrt{\beta^2 + 2\alpha \left[\sum_u W_{uv} \left(\hat{Y}_{ul}^{(t)} - \hat{Y}_{vl}^{(t)} \right)^2 + S_{v,v} \left(\hat{Y}_{vl}^{(t)} - Y_{vl} \right)^2 \right]}. \quad (3.7)
$$

From (3.6) we observe that the label scores at iteration $(t + 1)$ is a weighted combination of label scores from neighboring nodes from the previous iteration, and the seed label scores on the given node, if any. We observe that in contrast to other graph SSL algorithms, label scores are multiplied by a correcting factor which is inversely proportional to the label uncertainty on the node. Similarly from (3.7), we observe that uncertainty σ_{vl} for label l on node v, is directly proportional to the pairwise weighted disagreement in the score for the same label in the immediate neighborhood of node v. Thereby, for a given label, the higher the disagreement among neighbors, the lower the confidence for that label on the given node. Please note that this is irrespective of the degree of the node.

Since these updates only rely on label (and confidence) information from the current node and its immediate neighborhood, the updates can be easily parallelized, thereby enabling TACO to scale to large graphs. Orbach and Crammer [2012] also consider the case for non-diagonal, full Σ_v. However, they do not find any evidence that full confidence matrices improve performance over the diagonal approximations in practice.

3.2.8 INFORMATION REGULARIZATION

Information Regularization (IR) is a general framework for SSL. IR was first introduced by [Szummer and Jaakkola, 2001] and later refined in Corduneanu [2006] and Corduneanu and Jaakkola [2003]. The application of IR requires that one can define (possibly overlapping) regions over the input data. The IR objective prefers solutions in which samples occurring within a region have the same label. For example, in a document classification setting, all documents that have a particular word or set of words in the vocabulary may form a region. Yet another region may consist of all articles that were published close to a particular point in time, i.e., articles that are written close in time may have a similar label. It is important to note that the regions are defined based on domain knowledge about the problem and do not require labeled data.

Given a set of labeled and unlabeled examples, let $R \in \mathscr{R}$ denote a region where $\mathscr{R} = \{R_1, \ldots, R_{|\mathscr{R}|}\}$. Each input example, $\mathbf{x}_i \in \mathcal{D}$, $1 \leq i \leq n$, may be a member of multiple regions. The weight of $\mathbf{x}_i$ within a region is given by $p(\mathbf{x}_i|R)$ and $\sum_{\mathbf{x} \in \mathcal{D}} p(\mathbf{x}|R) = 1$. A region R may be formally defined as

$$R = \{\mathbf{x} \in \mathcal{D} \mid p(\mathbf{x}|R) > 0\}. \tag{3.8}$$

Each region is associated with a prior $p(R)$. The above probabilities are defined in a manner such that

$$\sum_{R \in \mathscr{R}} p(\mathbf{x}|R)p(R) = \sum_{R \in \mathscr{R}} p(\mathbf{x}, R) = \frac{1}{n}, \ \forall \ \mathbf{x} \in \mathcal{D}. \tag{3.9}$$

As a result, each sample, $\mathbf{x}_i \in \mathcal{D}$, is equally likely to be selected. Let $q(\mathbb{Y}|R)$ represent the distribution over the labels for a given region R. Let $r_i(\mathbb{Y})$ represent the seed label distribution for an instance from the supervised portion of the training data (i.e., $1 \leq i \leq n_l$). $r_i(\mathbb{Y})$ can be derived from the labels in a number of ways. For example, if y_i is the single supervised label for input $\mathbf{x}_i$, $1 \leq i \leq n_l$, then $r_i(y) = \delta(y = y_i)$, which means that r_i gives unity probability for y equaling the label y_i. The IR objective is given by:

$$\mathbf{C}^{(IR)}(p, q) = \sum_{i=1}^{n_l} \text{KL}(r_i(\mathbb{Y})||p(\mathbb{Y}_i|\mathbf{x}_i)) + \mu \sum_{R \in \mathscr{R}} p(R) \sum_{\mathbf{x} \in R} p(\mathbf{x}|R) \, \text{KL}(p(\mathbb{Y}|\mathbf{x})||q(\mathbb{Y}|R)),$$

$$\tag{3.10}$$

where $\text{KL}(r_i(\mathbb{Y})||p(\mathbb{Y}_i|\mathbf{x}_i)) = \sum_{y \in \mathcal{Y}} r_i(y) \log \frac{r_i(y)}{p(y|\mathbf{x}_i)}$ is the Kullback-Liebler divergence (KLD) between $r_i(\mathbb{Y})$ and $p(\mathbb{Y}_i|\mathbf{x}_i)$, similarly defined for $\text{KL}(p(\mathbb{Y}|\mathbf{x})||q(\mathbb{Y}|R))$, and μ is a hyperparameter that determines the relative importance of the two terms in the objective.

The first term in the above objective requires that final solution respect the labeled data but does not require that $p(y|\mathbf{x}_i) = r_i(y)$, as allowing for deviations from r_i can help especially with noisy labels [Bengio et al., 2007], or when the graph is extremely dense in certain regions. The second term is a smoothness penalty over both the labeled and unlabeled data defined in terms of the regions. It is given by

$$\sum_{R \in \mathscr{R}} p(R) \sum_{\mathbf{x} \in R} p(\mathbf{x}|R) \, \text{KL}(p(\mathbb{Y}|\mathbf{x})||q(\mathbb{Y}|R)) = \sum_{R \in \mathscr{R}} p(R) \sum_{\mathbf{x} \in R} \sum_{y \in \mathcal{Y}} p(\mathbf{x}|R) p(y|\mathbf{x}) \log \frac{p(y|\mathbf{x})}{q(y|R)}.$$

The above is minimized when each sample in a region is close to the "average" distribution for that region $q(y|R)$. In other words, if $p(y|\mathbf{x})$ and $q(y|R)$ are valid probability distributions then this term reaches its minimum value, 0, when all the labels for the points in that region are exactly the same. The IR regularizer is also intuitively appealing from an information theoretic standpoint. If a sender selects a $R \in \mathscr{R}$ according to $p(R)$, and a point within the region with probability $p(x|R)$, and subsequently a label y is sampled according to $p(y|\mathbf{x})$, then this label y is communicated to the receiver using a coding scheme tailored to the region based on $p(x|R)$ and $p(y|\mathbf{x})$. Minimizing the above term minimizes the rate of information needed to communicate with the receiver.

Thus, IR attempts to assign labels to all the unlabeled samples in way that is most consistent with the labeled examples while at the same time minimizing the disagreement of the inferred labels of the unlabeled samples within a region. Corduneanu and Jaakkola [2003] show that under certain conditions the IR objective is convex.

Lemma 3.1 [Corduneanu and Jaakkola, 2003] $\mathbf{C}^{(IR)}(p, q)$ *for $\mu > 0$ is strictly convex function of the conditionals $p(\mathbb{Y}|\mathbf{x})$ and $q(\mathbb{Y}|R)$ provided: (a) each $\mathbf{x}$ belongs to at least one region containing at least two points, and (b) the membership probabilities $p(\mathbf{x}|R)$ and $p(R)$ are all non-zero.*

Corduneanu and Jaakkola [2003] also show that the above objective may be solved using the *Blahut-Arimoto* algorithm [Vontobel, 2003]. The sequence of alternating updates (see Appendix C for more details about alternating minimization) to optimize the second term in $\mathbf{C}^{(IR)}(p, q)$, i.e., the minima for $\mathbf{C}^{(IR)}(p, q)$ for all $\mathbf{x}_i, n_l + 1 \leq i \leq n_l + n_u$, is given by

$$p^{(t)}(y|\mathbf{x}_i) \propto \exp\{t \sum_{R:\mathbf{x}_i \in R} p(R) p(\mathbf{x}_i|R) \log q^{(t-1)}(y|R)\} \tag{3.11}$$

$$q^{(t)}(y|R) = \sum_{\mathbf{x} \in R} p(\mathbf{x}|R) p^{(t)}(y|\mathbf{x}_i). \tag{3.12}$$

However, the update for $p(y|\mathbf{x}_i)$ for a labeled point, i.e., $1 \leq i \leq n_l$, does not admit a closed form solution. Tsuda [2005] have proposed a variant of the IR objective,

$$\mathbf{C}^{(IR')}(p, q) = \sum_{i=1}^{n_l} \text{KL}(r_i(\mathbb{Y})||p(\mathbb{Y}_i|\mathbf{x}_i)) + \mu \sum_{R \in \mathscr{R}} p(R) \sum_{\mathbf{x} \in R} \sum_{y \in \mathcal{Y}} p(\mathbf{x}|R) p(y|R) \log \frac{p(y|R)}{q(y|\mathbf{x})}$$

$$\tag{3.13}$$

and show that this is a dual of the original IR formulation. Notice the change in the direction of the KLD in the second term in the above formulation in comparison to the original IR formulation. $\mathbf{C}^{(IR')}(p, q)$ admits a closed form solution for both steps of the alternating minimization. While yielding similar results to that of the original IR objective (Equation 3.10), the above formulation is more efficient in practice.

It can be shown that IR subsumes many SSL algorithms such as harmonic graph regularization [Zhu and Ghahramani, 2002] and co-training [Blum and Chawla, 2001]. A graph may be represented by trivially defining each region, $R_i \in \mathcal{R}$, to represent an edge in the graph. It should be noted that while majority of the graph-based SSL algorithms only consider pairwise relationship between points, in the case of IR, one can represent constraints over larger number of points by defining a bipartite graph with input points on one side and appropriate regions on the other side. Corduneanu [2006] also suggest that IR is a special case of the popular clustering method *information bottleneck* [Tishby et al., 1999]. They also propose a simple message passing algorithm for solving the IR objective. The IR framework may be used to define both transductive and inductive learning algorithms.

3.2.9   MEASURE PROPAGATION

Measure propagation (MP) [Subramanya and Bilmes, 2010] is a probabilistic graph-based SSL algorithm. MP minimizes the following objective:

$$\mathbf{C}^{(MP)}(p) = \sum_{i=1}^{n_l} \mathrm{KL}\big(r_i(\mathbb{Y})||p(\mathbb{Y}_i|\mathbf{x}_i)\big) + \mu \sum_{i,j=1}^{n_l+n_u} W_{ij}\,\mathrm{KL}\big(p(\mathbb{Y}_i|\mathbf{x}_i)||p(\mathbb{Y}_i|\mathbf{x}_j)\big) +$$
$$\nu \sum_{i=1}^{n_l+n_u} \mathrm{KL}(p(\mathbb{Y}_i|\mathbf{x}_i)||u(\mathbb{Y})), \tag{3.14}$$

where μ and ν are hyperparameters which determine the relative importance of the different terms in the objective, $r_i(\mathbb{Y})$ is an encoding of the labeled data, and $u(\mathbb{Y})$ is a uniform distribution over the set of labels, i.e., $u(y) = \frac{1}{|\mathcal{Y}|} \ \forall\ y \in \mathcal{Y}$. Also, $p(y|\mathbf{x}_i)$ is the inferred label distribution for node $\mathbf{x}_i$ in the input graph and W_{ij} is the weight of the edge between nodes $\mathbf{x}_i$ and $\mathbf{x}_j$. The first term in $\mathbf{C}^{(MP)}(p)$ is similar to the first term in the IR objective $\mathbf{C}^{(IR)}(p)$ (3.10) in that it penalizes the solution $p(y|\mathbf{x}_i)(\forall 1 \le i \le n_l)$, when it is far away from the labeled training data $\mathcal{D}_l$. The second term of $\mathbf{C}^{(MP)}(p)$ penalizes a lack of consistency with the geometry of the data and can be seen as a graph regularizer. If W_{ij} is large, we prefer a solution in which $p(y|\mathbf{x}_i)$ and $p(y|\mathbf{x}_j)$ are close in the KLD sense. Finally, the last term encourages each $p(y|\mathbf{x}_i)$ to be close to the uniform distribution if not preferred to the contrary by the first two terms. Note that

$$\mathrm{KL}(p(\mathbb{Y}_i|\mathbf{x}_i)||u(\mathbb{Y})) = \sum_{y\in\mathcal{Y}} p(y|\mathbf{x}_i)\log p(y|\mathbf{x}_i) - \log\frac{1}{|\mathcal{Y}|}$$
$$\propto -H(\mathbb{Y}_i|\mathbf{x}_i),$$

where $H(\mathbb{Y}_i | X = \mathbf{x}_i)$ is the Shannon entropy of the conditional distribution. Thus we are maximizing the entropy of this distribution. This acts as a guard against degenerate solutions commonly encountered in graph-based SSL [Blum and Chawla, 2001, Joachims, 2003]. In essence, MP finds a labeling for the $\mathcal{D}_u$ that is consistent with both the labels provided in $\mathcal{D}_l$ and the geometry of the data induced by the graph.

Lemma 3.2 [Subramanya and Bilmes, 2010]   *If $\mu, \nu, W_{ij} \geq 0$ $(1 \leq i, j \leq n)$, then $\mathbf{C}^{(MP)}(p)$ is convex.*

As $\mathbf{C}^{(MP)}(p)$ is convex and the constraints are linear, it can be solved using convex programming [Bertsekas, 1999]. However, this optimization problem does not admit a closed form solution as it requires optimizing with respect to both variables in a KLD. Subramanya and Bilmes [2010] propose a numerical method based on the method of multipliers with a quadratic penalty to solve the above problem. But this has a number of drawbacks ranging from the introduction of a number of extraneous optimization related hyperparameters which can be hard to tune, to the lack of convergence guarantees. Subramanya and Bilmes [2010] also propose a reformulated version of the MP objective that lends itself to closed-form updates.

Re-formulated Objective   Consider the following:

$$\mathbf{C}^{(RMP)}(p,q) = \sum_{i=1}^{n_l} \mathrm{KL}\big(r_i(\mathbb{Y})||q(\mathbb{Y}_i|\mathbf{x}_i) + \mu \sum_{i,j=1}^{n_l+n_u} W'_{ij} \mathrm{KL}\big(p(\mathbb{Y}_i|\mathbf{x}_i)||q(\mathbb{Y}_i|\mathbf{x}_j)\big) +$$
$$\nu \sum_{i=1}^{n_l+n_u} \mathrm{KL}(p(\mathbb{Y}_i|\mathbf{x}_i)||u(\mathbb{Y})),$$

where for each vertex $\mathbf{x}_i$ in G, we have another distribution $q(\mathbb{Y}_i|\mathbf{x}_i)$ over the measurable space $(Y, \mathcal{Y})$, $W' = W + \alpha I_n$ $(\alpha \geq 0)$, $\Gamma'(\mathbf{x}_i) = \Gamma(\mathbf{x}_i) \cup \{\mathbf{x}_i\}$. Here, the $q(y|\mathbf{x}_i)$'s play a similar role as the $p(y|\mathbf{x}_i)$'s and can potentially be used to obtain a final classification result $(\mathrm{argmax}_y\, q(y|\mathbf{x}_i))$. Thus, it would seem that we now have two classification results for each sample—one the most likely assignment according to $p(y|\mathbf{x}_i)$ and another given by $q(y|\mathbf{x}_i)$. However, the above objective includes terms of the form $\alpha \times \mathrm{KL}(p(y|\mathbf{x}_i)||q(y|\mathbf{x}_i))$ which encourage $p(y|\mathbf{x}_i)$ to be close to $q(y|\mathbf{x}_i)$. Thus, α, which is a hyper-parameter, plays an important role in ensuring that $p(y|\mathbf{x}_i) = q(y|\mathbf{x}_i)$ $(\forall 1 \leq i \leq n)$. Thus, the second term in the above objective, while acting as a graph regularizer, also acts as a glue between the p's and q's. Subramanya and Bilmes [2010] show that the above objective is amenable to optimization using *alternating minimization* (see Appendix C for more details) and the sequence of updates is given by

$$p^{(t)}(y|\mathbf{x}_i) = \frac{\exp\{\frac{\mu}{\gamma_i} \sum_{j=1}^{n} W'_{ij} \log q^{(t-1)}(y|\mathbf{x}_j)\}}{\sum_{y \in \mathcal{y}} \exp\{\frac{\mu}{\gamma_i} \sum_{j=1}^{n} W'_{ij} \log q^{(t-1)}(y|\mathbf{x}_j)\}}$$
$$q^{(t)}(y|\mathbf{x}_i) = \frac{r_i(y)\delta(i \leq n_l) + \mu \sum_{j=1}^{n} W'_{ji}\, p^{(t)}(y|\mathbf{x}_j)}{\delta(i \leq n_l) + \mu \sum_{j=1}^{n} W'_{ji}},$$

where $\gamma_i = \nu + \mu \sum_{j=1}^{n} W'_{ij}$.

Subramanya and Bilmes [2010] prove that there exists a finite α so that the minimum of $\mathbf{C}^{(RMP)}(p, q)$ is the same as that of $\mathbf{C}^{(MP)}(p)$. However, they do not provide a recipe for finding such an α. Setting $\alpha = 1$ has been found to work well in practice.

Measure Propagation vs. Information Regularization   As has been noted above, IR is a general SSL framework which can also be applied to graph-based SSL problems. In comparison, MP is a graph-based SSL algorithm. While one of the terms in the IR objective can be optimized using a alternative minimization based approach, the other does not admit a closed form solution. This is a serious practical drawback especially in the case of large data sets. MP, on the other hand, admits a closed form solution thus making it easy to scale to different problem settings.

All the graph SSL methods discussed so far (except for IR) are *transductive* in nature. Thus, in theory they cannot be applied to previously unseen data instances. In practice however, one can use nearest neighbor methods to apply transductive methods to inductive settings. One such method is the *Nadaraya-Watson estimator*. Given a new sample $\mathbf{x}$, its label can be estimated using

$$\hat{p}(y) = \frac{\sum_{j \in \Gamma(\mathbf{x})} \text{sim}(\mathbf{x}, \mathbf{x}_j) p^*(y | \mathbf{x}_j)}{\sum_{j \in \Gamma(\mathbf{x})} \text{sim}(\mathbf{x}, \mathbf{x}_j)}, \tag{3.15}$$

where $\Gamma(x)$ are the set of nearest neighbors of $\mathbf{x}$ in the graph G. $p^*(y | \mathbf{x}_j)$ are the converged values of the label distributions at node j. As it can be seen, one of the downsides of this approach is that, unlike inductive methods, it requires that one have access to the entire training dataset (both labeled and unlabeled).

3.3   INDUCTIVE METHODS

In this section, we present *inductive* methods which use the graph structure to estimate a function which can then be applied to new data instances. We note that while several transductive graph SSL methods exist, inductive methods are quite limited in number, with Manifold Regularization [Belkin et al., 2005] the representative example of this class of methods. While all the transductive methods discussed so far are capable of handling multiple classes, the inductive methods discussed in this section mostly only work in the binary setting. These methods estimate a function $f : \mathcal{X} \to \mathcal{R}$, whose output score may be turned into binary classification decision by thresholding appropriately. Although these binary classifiers may be applied in a multi-class setting by estimating a separate function for each class, and using the resulting classifiers in a one-vs.-all setting, that is different from intrinsic support for multiple classes in the transductive methods such as MP and MAD.

3.3.1   MANIFOLD REGULARIZATION

Manifold regularization [Belkin et al., 2005] is based on the assumption that $p(y|x)$ varies smoothly with $p(x)$, i.e., if two samples x_1 and x_2 are *close* then their conditional distributions $p(y|x_1)$ and $p(y|x_2)$ are similar. Let $f : \mathcal{X} \to \mathcal{R}$ be a function that maps the input to the output and $\mathcal{L}(x, y, f)$ represents the loss incurred from mapping x to $f(x)$ when the expected output is y.

For a Mercer kernel $K : \mathcal{X} \times \mathcal{X} \to \mathcal{R}$, there is an associated Reproducing Kernel Hilbert Space (RKHS), $\mathcal{H}_K$ of functions, $f : \mathcal{X} \to \mathcal{R}$ with the corresponding norm $||.||_K$. Consider

$$f^*(x) = \underset{f \in \mathcal{H}_K}{\text{argmin}} \frac{1}{n_l} \sum_{i=1}^{n_l} \mathcal{L}(x_i, y_i, f) + \lambda_a \parallel f \parallel_K^2 + \lambda_b \parallel f \parallel_I^2, \qquad (3.16)$$

where $\mathcal{L}$ is the loss function defined over the labeled examples. Here, $\parallel f \parallel_K^2$ is a "standard" regularizer that ensures the smoothness of possible solutions, while $\parallel f \parallel_I^2$ is a smoothness penalty on the probability distribution $p(x)$. λ_a and λ_b are hyperparameters that determine the relative importance of these penalties. In almost all practical situations, the marginal $p(x)$ is not known but can be estimated empirically using the unlabeled data. Belkin et al. [2005] show that in cases where the unlabeled data lies on a manifold $\mathcal{M} \subset \mathcal{R}^n$, then $\parallel f \parallel_I^2$ may be approximated by using the graph Laplacian over the labeled and unlabeled data. Thus, the objective in the case of manifold regularization is given by

$$f^*(x) = \underset{f \in \mathcal{H}_K}{\text{argmin}} \frac{1}{n_l} \sum_{i=1}^{n_l} \mathcal{L}(x_i, y_i, f) + \lambda_a \parallel f \parallel_K^2 + \frac{\lambda_b}{(n_l + n_u)^2} \sum_{i,j=1}^{n} W_{ij}(f(x_i) - f(x_j))^2,$$

where W_{ij} are the edge weights in the graph. We note that $\sum_{i,j=1}^{n} W_{ij}(f(x_i) - f(x_j))^2 = \mathbf{f}^T L \mathbf{f}$ where $\mathbf{f} = [f(x_1), \ldots, f(x_n)]^T$ and $L = D - W$ is the unnormalized graph Laplacian. Hence, the MR objective above may be equivalently written as

$$f^*(x) = \underset{f \in \mathcal{H}_K}{\text{argmin}} \frac{1}{n_l} \sum_{i=1}^{n_l} \mathcal{L}(x_i, y_i, f) + \lambda_a \parallel f \parallel_K^2 + \frac{\lambda_b}{(n_l + n_u)^2} \mathbf{f}^T L \mathbf{f}$$

Theorem 3.3 [Belkin et al., 2005]. *The minimizer of the above optimization problem (3.16) admits an expansion*

$$f^*(x) = \sum_{i=1}^{n} \alpha_i K(x, x_i)$$

in terms of the labeled and unlabeled examples.

Belkin et al. [2005] suggest that using the *normalized Laplacian*, $\tilde{L} = D^{-1/2} L D^{1/2}$ has advantages both from theoretical and practical perspectives. Two popular variants of the manifold regularization arise from using squared error and hinge loss for the loss function. They are referred to as *Laplacian Regularized Least Squares* (LapRLS) and *Laplacian Support Vector Machine* (LapSVM), respectively. We will review these two methods next.

Laplacian Regularized Least Squares   The optimization problem for Laplacian Regularized Least Squares (LapRLS) is given by

$$\underset{f \in \mathcal{H}_K}{\operatorname{argmin}} \frac{1}{n_l} \sum_{i=1}^{n_l} (y_i - f(x_i))^2 + \lambda_a \parallel f \parallel_K^2 + \frac{\lambda_b}{(n_l + n_u)^2} \mathbf{f}^T L \mathbf{f}.$$

It can be shown that the representer theorem applies in this case and so $f^*(x) = \sum_{i=1}^{n} \alpha_i^* K(x, x_i)$ and that the final solution is given by

$$\alpha^* = \left(SK + \lambda_a \times n_l \times I + \frac{\lambda_b \times n_l}{(n_l + n_u)^2} LK \right)^{-1} \mathbf{y},$$

where $\mathbf{y} = [y_1, \ldots, y_l]$ and $S \in \mathcal{R}^{n \times n}$ is the seed indicator diagonal matrix with the first n_l diagonal entities set to 1 and all other entries set to 0. Note that the solution requires inversion of a matrix of size $n \times n$.

Laplacian Support Vector Machines   Using a hinge loss instead of the square loss, Laplacian Support Vector Machines (LapSVM) solves the following optimization problem:

$$\underset{f \in \mathcal{H}_K}{\operatorname{argmin}} \frac{1}{n_l} \sum_{i=1}^{n_l} |1 - y_i f(x_i)|_+ + \lambda_a \parallel f \parallel_K^2 + \frac{\lambda_b}{(n_l + n_u)^2} \times \mathbf{f}^T L \mathbf{f}.$$

Here, $|1 - yf(x)|_+ = \max(0, 1 - yf(x))$ is the hinge loss defined under the assumption $y_i \in \{-1, +1\}$. Again, the solution in the case is given by $f^*(x) = \sum_{i=1}^{n} \alpha_i^* K(x, x_i)$. Unlike LapRLS, there is no closed form solution for the α_i's in LapSVM, and they are instead estimated using standard SVM solvers. For more details, see Belkin et al. [2005].

3.4   RESULTS ON BENCHMARK SSL DATA SETS

In this section we compare the performance of some of the SSL algorithms presented in this chapter. We choose five benchmark SSL datasets from Chapelle et al. [2007]. They involve a variety of tasks—Digit1 consists of transformed versions of the digit "1" chosen so that the points like on a low-dimensional manifold. USPS is a handwritten digit recognition task with classes "2" and "5" assigned the positive label while all other classes were assigned the negative label. COIL or the Columbia object image library consists of a set of color images of 100 different objects taken from different angles. BCI is data collected from a research effort aimed at building brain-computer interfaces. It consists of appropriately transformed EEG data. Finally, text consists of five groups from the Newsgroups dataset and the goal is to classify the *ibm* category vs. the rest.

Table 3.1: Test errors (lower is better) of various algorithms on the benchmarks datasets used in [Chapelle et al., 2007], available for download at `http://olivier.chapelle.cc/ssl-book/ben chmarks.html`. For all experiments, 100 labeled instances were used, and model selection was performed on test data (as suggested in Chapelle et al. [2007])

Approach	Digit1	USPS	COIL	BCI	Text
1-NN	6.1	7.6	23.3	44.8	30.8
Linear TSVM [Joachims, 1999]	18.0	21.2	-	42.7	22.3
SGT [Joachims, 2003]	2.6	6.8	-	45.0	23.1
GRF [Zhu et al., 2003]	2.5	5.7	11.0	46.6	26.8
Adsorption [Baluja et al., 2008]	5.9	3.4	10.5	46.5	26.6
MAD [Talukdar and Crammer, 2009]	2.9	4.9	9.9	45.5	26.6
QC [Bengio et al., 2007]	3.2	6.4	10.0	46.2	25.7
IR [Corduneanu and Jaakkola, 2003]	2.4	5.1	11.5	47.5	-
MP [Subramanya and Bilmes, 2010]	2.6	3.8	9.8	46.0	22.9
LapRLS [Belkin et al., 2005]	2.9	4.7	11.9	31.4	23.6
LapSVM [Belkin et al., 2005]	3.1	4.7	13.2	28.6	23.0

COIL is the only multiclass dataset ($m = 6$) while all others are binary. The features in the case of the Text dataset are sparse and high-dimensional ($d = 11,600$). The advantage of these datasets is that feature vectors and transduction splits are available making it easy to compare performance[1]. More details about these datasets can be found in Chapelle et al. [2007].

Results obtained from the various algorithms are in Table 3.1. 1-NN represents classification using 1-nearest neighbor and is fully supervised. All the other algorithms are semi-supervised. Among the SSL algorithms used, linear TSVM [Joachims, 1999] is the only non-graph-based SSL algorithm. In order to make it easier to interpret the results, Chapelle et al. [2007] propose splitting the datasets into two categories: those that are more *manifold-like*, in that the data most likely lies near a low-dimensional manifold and those that are *cluster-like*, i.e., the data clusters such that two classes do not share the same cluster. They argue that Digit1, USPS, COIL and BCI are manifold-like while Text is cluster-like. Linear TSVM and IR are cluster-based algorithms and thus expected to perform better on cluster-like data, while the remaining SSL algorithms are manifold-based, i.e., they define their loss with respect to a manifold (the graph) and so we expect them to do better on manifold-like data.

All experiments are in the transductive setting. In order to judge the potential of a method, model selection is performed on the test data. As can be seen, no single algorithm performs the best for all settings. This implies that the appropriate algorithm needs to be carefully selected based on the nature of the data. It can also be seen that most of the graph-based SSL algorithms outperform linear TSVMs on a majority of the datasets. Results for the BCI dataset, where the

---

[1]These datasets can be downloaded from `http://olivier.chapelle.cc/ssl-book/benchmarks.html`.

supervised 1-NN performs as well (and in some cases better) than all the other SSL algorithms, show that one cannot always get improved performance by making use using unlabeled data.

3.5   CONCLUSIONS

Table 3.2: Comparison of the different SSL algorithms presented in this chapter along various dimensions

Approach	Type	Loss	Discount Nodes	Probabilistic
Mincut	Transductive	L1 norm	No	No
SGT	Transductive	L1 norm	No	No
GRF	Transductive	Squared Error	No	No
Adsorption	Transductive	Squared Error	Yes	No
MAD	Transductive	Squared Error	Yes	No
QC	Transductive	Squared Error	No	No
IR	Transductive	KL-Divergence	No	Yes
MP	Transductive	KL-Divergence	No	Yes
LapRLS	Inductive	Squared Error	No	No
LapSVM	Inductive	Squared Error	No	No

We have presented a number of graph-based SSL algorithms in this chapter. Table 3.2 compares these algorithms along a number of dimensions. We started with early approaches which cast graph-based SSL as a graph cut problem [Blum and Chawla, 2001]. These approaches only provide a hard classification result without any indication of confidence and sometimes lead to degenerate solutions, i.e., solutions in which all the unlabeled samples are classified as belonging to a single class. Spectral graph transduction (SGT) [Joachims, 2003] addresses the later concern by using a norm-cut based objective. We then presented the seminal work of Zhu et al. [2003] who proposed a convex objective for graph-based SSL. The resulting algorithm is popularly referred to as *label propagation* (LP).

We then presented the Adsorption algorithm [Baluja et al., 2008] which takes a random walk route to solve the graph-based SSL problem. Talukdar and Crammer [2009] show that there is no function whose local minima is the same as the output of Adsorption and proposed a modified version called *Modified Adsortion* or MAD which can be expressed as a convex optimization problem. Both of the above algorithms use a fixed node discounting strategy that is determined *a priori*, while Transduction with Confidence (TACO) [Orbach and Crammer, 2012] makes use of a data-driven strategy to discount nodes. TACO takes into consideration the agreement of label assignment between a node and its neighbors to estimate confidence of label assignments.

Information regularization (IR) [Corduneanu, 2006] subsumes many SSL algorithms such as co-training and can be used to solve graph-based SSL problems. One downside of IR is that one of the terms in the objective does not admit a closed form solution. Measure propagation

(MP) [Subramanya and Bilmes, 2010], on the other hand, can be optimized using alternating minimization and each of the steps admits a closed form solution. Both MP and IR are probabilistic methods and extend naturally to multi-class SSL problems. Further, MP and IR both define their loss functions using the Kullback-Liebler divergences while majority of the other approaches use squared error. Finally, all of the above approaches (except IR) are trasductive; Manifold regularization [Belkin et al., 2005] is an inductive method.

The objective used in the case of a majority of the graph-based SSL algorithms has a common form:

$$\text{objective} = \boxed{\text{be close to labeled data}} + \boxed{\text{be smooth over the graph}} + \boxed{\text{regularizer}}$$

where the first term encourages the final solution to be close to the labeled data. While approaches such as Gaussian random field (GRF) [Zhu et al., 2003] require that the final solution for the labeled nodes exactly match the initial seed labels, others such as MAD, MP, and QC allow the final solution to deviate from the initial seed labels. This has been shown to make these algorithms resilient to label noise. The second term above prefers solutions which vary smoothly over the graph. This implies that two nodes connected by a highly weighted edge should get similar (or ideally the same) label scores. The third term is a regularizer that controls the complexity of the final solution. In the case of QC, MAD and MP, the third term is used as a guard against degenerate solutions.

Each of the algorithms presented in this chapter have their pros and cons; the choice of the algorithm should be based on the characteristics of the problems. Some factors to consider include, how much labeled data does one have, are the labels zero entropy or are they distributions, are the labels noisy, how much unlabeled data does one have, what is the nature of the data (e.g., is it expected to lie near a low dimensional manifold) and so on. Even though there has been a large amount of work in the area of inference in graph-based SSL, there are still a number of open problem which we will discuss in Chapter 6.

CHAPTER 4

Scalability

Thus far we have seen how to construct a graph and subsequently use it to infer labels for the unlabeled samples. Scaling up both of these steps will be the focus of this chapter. We first present some algorithms for constructing graphs over a large number of samples followed by inference in a number of parallel architectures including shared-memory symmetric multi-processors (SMPs) and distributed computers.

Why scale graph-based SSL?   As we have explained previously, SSL is based on the premise that small amounts of labeled data combined with relatively large amounts of unlalabeled data can lead to much improved performance.[1] But can large amounts of unlabeled data be easily obtained? On the Internet, about 1.6 billion blog posts, 60 billion emails, 2 million photos, and 200,000 videos are created every day [Tomkins, 2008]. As the above estimates are a few years old, the numbers today are probably at least one-order-of-magnitude larger. As a result it has become critical that an algorithm and in particular a SSL algorithm scale easily to large data problems.

4.1   LARGE-SCALE GRAPH CONSTRUCTION

As noted in Chapter 2, graph construction is an important step in the application of any graph-based SSL algorithm. The graph structure determines the neighborhood structure and strength of association between a pair of nodes. These association strengths determine how information propagates from one node to another during execution of the graph SSL algorithm. Thus, improper choice of this graph can lead to severe degradation of performance, with possibility of degenerate solutions [Blum and Chawla, 2001, Joachims, 2003].

Most of the widely used approaches for graph construction (e.g., k-Nearest Neighbors) are quadratic (or in some cases more expensive) in the number of examples. As a result they do not scale well when the number of data samples increase. Furthermore, many of the other graph construction approaches also involve choosing one or more hyperparameters which can be difficult in large data scenarios. Next we present some approaches for constructing graphs for solving SSL problems with a large number of samples.

4.1.1   APPROXIMATE NEAREST NEIGHBOR

k-Nearest Neighbor (kNN) remains one of the more popular graph construction techniques (Section 2.2.1). Conventional brute-force approach to constructing a kNN graph is $O(n^2)$ in

---

[1]Nadler et al. [2010] show that for certain graph-based SSL approaches, large amounts of unlabeled data can hurt performance.

the number of samples and thus does not scale well even in the case of moderately sized problems (e.g., $n = 10$ million). The topic of faster but approximate nearest neighbor search has been well researched. *kd-trees* [Bentley, 1975, Friedman et al., 1977] are a class of data structures that allow $O(\log n)$ nearest neighbor search thereby rendering the graph construction problem $O(n \log n)$. One downside of kd-trees is that they do not perform well in the case of sparse or high-dimensional feature vectors. Most natural language processing applications involve the use of sparse feature vectors.

A modified kd-tree data structure is presented in Arya and Mount [1993] and Arya et al. [1998], which can be used to query nearest neighbors. This data structure is implemented in the Approximate Nearest Neighbor (ANN) library (see `http://www.cs.umd.edu/~mount/ANN`), and has been shown to work well for sparse and/or high-dimensional features. It satisfies an approximation guarantee of

$$\frac{\text{sim}(\mathbf{x}_i, \text{N}(\mathbf{x}_i, 0))}{\text{sim}(\mathbf{x}_i, \text{N}(\mathbf{x}_i, \epsilon))} \leq 1 + \epsilon,$$

where ϵ is a pre-specified error tolerance level, and $\text{N}(\mathbf{x}_i, \epsilon)$ is the approximate NN returned by the query process. We note that improved query speed may be achieved using higher ϵ, although often at the cost of accuracy. Please refer to Arya and Mount [1993] and Arya et al. [1998] for more details.

Locality sensitive hashing (LSH) can also be used to perform approximate nearest neighbor search. It is based on the idea of using a hash function that maps two points that are close in the input space to the same bucket. In other words, the key idea is to choose a hash function such that the probability of collision is larger for objects that are close to each other than for those that are far apart. In order to improve the reliability, one often uses a number of hash functions chosen randomly from a family of hash functions, $\mathcal{H}$. Given a new sample, it is first hashed using all the functions and all the points that also hashed into the same respective buckets are retrieved and the smallest distance between the input sample and those points is computed. More details can be obtained in Andoni and Indyk [2008].

4.1.2   OTHER METHODS

Delalleau et al. [2005] propose creating a small graph with a subset of the unlabeled data thereby enabling fast computation. We are not aware, however, of a published principled algorithm to choose such a subset. Garcke and Griebel [2005] propose the use of sparse grid for semi-supervised learning. The idea was to approximate the function space with a finite basis with sparse grids. While their approach scales linearly in the number of samples, in practice, it only works for relatively low-dimensional (< 20) data.

Liu et al. [2010] propose a scalable graph construction algorithm based on anchor points. The idea here is to use a small set of n' ($\ll n$) nodes to approximate neighborhood structure. They also show how the anchor points can be used for fast inference. A sketch-based approximate

---

Algorithm 4: Graph Node Ordering Algorithm Pseudocode (SMP Case)

Input: Graph $G = (V, E)$
Result: Node ordering
Select an arbitrary node v;
while *There are unselected nodes remaining* **do**
> Select an unselected $v' \in \Gamma(\Gamma(v))$ that maximizes $|\Gamma(v) \cap \Gamma(v')|$;
> If the intersection is empty, select an arbitrary unselected v';
> Mark v' as selected v' is next node in the order $v \leftarrow v'$;

---

counting scheme to construct graphs from large number of samples is presented in Goyal et al. [2012].

4.2   LARGE-SCALE INFERENCE

As we discussed in Chapter 3, given a graph, the next step in solving a SSL problem using graph-based methods is inferring the labels for the unlabeled samples. Further, as we saw in Chapter 3, a common feature of the inference in most graph-based algorithms is that their optimization can be expressed as simple and efficient messages passed along edges of the graph. As a result, the manner in which the graph is organized plays an important role in determining the speed of inference. Graph partitioning is not only important in distributed inference scenarios but also in the case where the entire graph can be loaded into the memory of a single machine.

4.2.1   GRAPH PARTITIONING

In this section, we first review techniques to partition the graph keeping properties of subsequent inference in mind. In Section 4.2.1, we consider the problem of partitioning graphs in a shared-memory symmetric multi-processor (SMP) setting (i.e., cases where the entire graph can be loaded into a single machine) and then present partitioning in a distributed setting.

Shared-Memory Symmetric Multi-Processor (SMP)   In a SMP, all processors share a common main memory and are often cache coherent. Such settings are of importance in today's computing environments where multiple-cores are available in a single package. As we observed in Chapter 3, iterative updates of many graph SSL algorithms may be implemented in parallel using a vertex-based partitioning. So, given an SMP with T threads, it is natural to partition a graph with n nodes into $\frac{n}{T} = \frac{n_l + n_u}{T}$ subsets and process each subset in parallel. In other words, if we want to perform inference using T threads, we would like to generate the following:

- a partition of the nodes V of the graph into $\frac{n}{T}$ subsets $\{V_1, \dots, V_{\frac{n}{T}}\}$, with each $|V_i| \approx T$, and process all nodes in a single subset in parallel across all threads T, and

- an ordering, π, of the partitions generated above so that nodes in subset V_{π_i} are processed before the nodes in partition $V_{\pi_{i+1}}$.

Let $\mathbf{x}_t$ be the node currently processed in thread t and so $\cup_{1 \leq t \leq T} \{\mathbf{x}_t\}$ is the set of nodes currently being processed across the T threads. Most iterative graph SSL algorithms require information only from the immediate neighborhood of a given node to update the label distribution at that node. Let us define *working set* as the union of neighborhoods of all nodes currently being processed, i.e., $\cup_{1 \leq t \leq T} \Gamma(\mathbf{x}_t)$ where $\Gamma(\mathbf{x}_t)$ is the set of neighbors of node $\mathbf{x}_t$ in the graph. Ideally, we would like to make this working set as small as possible so that it can be fit in the shared cache. However, this becomes challenging as T increases since the working set monotonically increases with increasing T. In addition, we would like to make sure neighborhood overlap among nodes of successive partitions V_{π_i} and $V_{\pi_{i+1}}$ is maximized. This ensures that neighbors of nodes in $V_{\pi_{i+1}}$ are already pre-fetched by the time their processing starts.

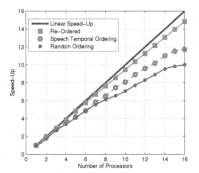 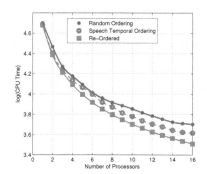

Figure 4.1: Speed-up and runtime against number of processors (T) when Measure Propagation (MP) is run over a graph constructed using the Switchboard corpus [Godfrey et al., 1992], with and without node reordering (Algorithm 4). We observe that the node reordering algorithm achieves close to linear speedup, and significantly reduced runtimes. Figure is reproduced from Bilmes and Subramanya [2011].

Algorithm 4 was proposed by Bilmes and Subramanya [2011] to reorder the graph to satisfy the above requirements. The algorithm favors orderings where successive nodes have significant neighborhood overlap. Given the current node $\mathbf{x}_v$, the algorithm selects the next node in the ordering, $\mathbf{x}_{v'}$, such that the neighborhood overlap between the two nodes is maximum. The key insight is that, for an undirected graph, a search over $\Gamma(\Gamma(\mathbf{x}_v))$, the set of neighbors of the neighbors of $\mathbf{x}_v$ is sufficient, as any node outside of this set has no neighbor overlap with $\mathbf{x}_v$ anyways and hence will not be selected. We note that Algorithm 4 does not take the value of T into consideration during order estimation. Once the ordering is determined, successive chunks of T nodes are considered a subset, and processed in parallel. This not only ensures high neighborhood overlap among nodes in a single subset, but also between consecutive subsets—the two desirable properties stated above.

Bilmes and Subramanya [2011] applied Algorithm 4 to a graph with 120 million nodes constructed from the Switchboard corpus (a speech dataset) [Godfrey et al., 1992]. The number of threads, T, was varied from 1–16, and Measure Propagation (MP) was used as the inference algorithm in all these experiments. Experimental results comparing speedup and runtimes with and without the reordering algorithm are presented in Figure 4.1. Experiments were repeated 10 times and the minimum runtime across all these runs was considered. The speedup with T threads is defined as the fraction of runtime with T threads compared to single thread runtime. From Figure 4.1, we observe that the node ordering algorithm achieves close to linear speedup, and runtime improves even for the $T = 1$ case. For fair comparison, CPU time required for node reordering was included in all reported runtimes of experiments involving the node reordering algorithm. These results demonstrate benefits to be had through careful partitioning of large graphs before inference. The results also show that the speedups achieved by using the proposed reordering algorithm are significantly larger than those obtained in the case of using a random ordering where the consecutive node is chosen randomly from the set of remaining nodes. Please refer to Bilmes and Subramanya [2011] for more details.

Although the experiments reported here are using one specific graph SSL algorithm, the idea of graph partitioning and reordering is general and any graph SSL algorithm that exploits neighbor structure for inference may benefit from it. Some recent work in this area is presented in Ugander and Backstrom [2013]. Next we present partitioning in a distributed computing environment.

Distributed computing environments   provide an alternative to SMPs, especially when one is interested in utilizing inexpensive and massive parallelism. Unlike SMPs, distributed computing environments usually consist of tens of thousands of computing nodes without any shared memory, and communication between nodes is relatively expensive and is usually achieved using some form of messaging library.

As before, in the distributed computer setup here, we are interested in partitioning the nodes V of the graph $G = (V, E)$ into T subsets $V = V_1 \cup V_2 \cup \ldots \cup V_T$ with $V_i \cap V_j = \emptyset \; \forall 1 \leq i, j \leq T, i \neq j$. All nodes from a single subset are processed in one of the T computing processors. Within each subset, which is processed on a single compute node, we would still like to maximize the neighborhood overlap among the nodes (using Algorithm 4). However, the similarities with SMP end here as a distributed computer has several unique characteristics. Since, message passing-based communication between processors is relatively expensive, we would like to minimize inter-processor communication. Given this, we would like to minimize the neighbor overlap between two subsets assigned to two different processors. In other words, we would like to minimize $|\Gamma(V_i) \cap \Gamma(V_j)| \; (i \neq j)$, where V_i and V_j are two subsets assigned to two different processors. This is in contrast to the SMP setting where we wanted a large degree of neighborhood overlap across subsets, especially the ones ordered consecutively.

Algorithm 5: Graph Node Ordering Algorithm for a Distributed Computer

---

Input: A Graph $G = (V, E)$ with $n = |V|$ nodes. Parameter T indicating the number
of compute nodes. A positive integer threshold τ.
Result: A node ordering, by when they are marked.
Select an arbitrary node v;
$i \leftarrow 0$;
while *There are unselected nodes remaining* **do**

    if $\min_\ell |i - \ell \times \frac{n}{T}| < \tau$ **then**   `// near a transition`
        Select uniformly at random any unselected node v' ;
    else   `// not near a transition`
        Select an unselected $v' \in \Gamma(\Gamma(v))$ that maximizes $|\Gamma(v) \cap \Gamma(v')|$. If the
        intersection is empty, select an arbitrary unselected v';

    Mark v' as selected.;   `// v' is next node in the order`
    $v \leftarrow v'$;
    $i \leftarrow i + 1$;

foreach ℓ **do** `// randomly scatter boundary nodes to internal locations`
    Define segment ℓ boundary node indices as
    $B_\ell = \{i : 0 \le i - \ell \times \frac{n}{T} < \tau \text{ or } 0 \le (\ell + 1) \times \frac{n}{T} - i < \tau\}$;
    foreach $i \in B_\ell$ **do**
        Insert node i uniformly at random between nodes $\ell \times \frac{n}{T} + \tau$ and
        $(\ell + 1) \times \frac{n}{T} - \tau$;

The optimization problem above may be posed as a T-way-cut problem, whose exact solution would have given a portioning of the nodes in the graph, where within set neighborhood overlap is maximized, while minimizing across set neighborhood overlap. Unfortunately, this problem is known to be NP-complete [Vazirani, 2001], and even though a $(2 - \frac{2}{T})$ approximation exists. This approximation algorithm uses a Gomory-Hu tree [Vazirani, 2001] which requires computing $(n - 1)$ (s, t)-cuts, making it impractical for very large-scale problems with millions of nodes ($n \gg 1$). The requirement for a normalized (balanced) partition, i.e., requiring all subsets to be of equal size, increases complexity. Solving even for normalized 2-partitions is NP-complete [Shi and Malik, 2000].

Bilmes and Subramanya [2011] propose a modification of the SMP heuristic (Algorithm 4) keeping in mind the challenges of the distributed setting. The idea is to adopt a hybrid strategy that chooses from the following options: (1) choose the next node from the neighbors' of neighbors, while maximizing neighborhood overlap; and (2) choose the next node uniformly at random. Assuming a normalized balance, the subset divisions occur at nodes $\frac{q \times n}{T}, 0 \le q \le T$. Let i be the current position in the ordering. If i is far away from any transition node, i.e., i is more than τ

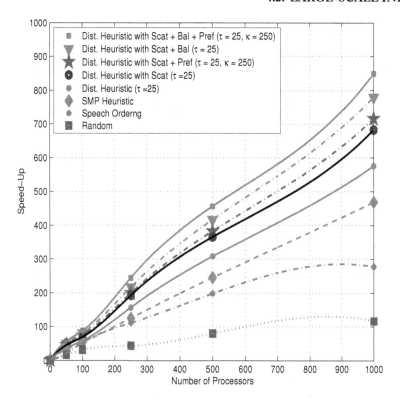

Figure 4.2: Speedup vs. number of processors for inference the Switchboard graph with and without the node ordering algorithm (Algorithm 5) in a distributed environment. Figure is reproduced from Bilmes and Subramanya [2011].

nodes away from the nearest transition node, then the degree of neighbor overlap among nodes immediately preceding and succeeding i in the ordering should be high. However, if i is within τ nodes of some transition node, then i is in a boundary region, and the node at that position should be chosen at random so that the neighborhood overlap is minimized. This strategy is implemented in the first *while* loop of Algorithm 5.

Bilmes and Subramanya [2011] implement this algorithm on a 1000 node distributed computer (cluster). Experimental results comparing different algorithms are presented in Figure 4.2. First, we observe that the random ordering performs quite poorly. Further improvement is achieved when the SMP heuristic (Algorithm 4) is applied in the distributed computer. Please note that the SMP heuristic doesn't make any effort to ensure $|\Gamma(V_i) \cap \Gamma(V_j)|$ is small for $i \neq j$. Further speedups are observed when only lines 3–10 of Algorithm 5 are used, marked as "Dist. Heuristic" (with $\tau = 25$) in Figure 4.2. Lines 11–14 in Algorithm 5 randomly *scatter* boundary nodes into non-boundary segments. This results in improved performance which is marked "Dist. Heuristic with Scat + Pref." in Figure 4.2.

Once scattering has been performed, we know a priori which nodes are scattered, and thereby which nodes are unlikely to have their neighbors pre-fetched. We can exploit this information and pre-fetch neighbors of scattered nodes. This results in further improvements in efficient which is marked as "Dist. Heuristic with Scat + Pref." in the figure. Finally, pre-fetching and load balancing may be used together to improve performance even further, which is marked as "Dist. Heuristic with Scat + Bal + Pref." in Figure 4.2.

We note that these heuristics are fairly generic, which have resulted in performance improvement even when applied to a distributed computer without any prior knowledge of the network topology. Further gains may be possible with more knowledge about latency, bandwidth limitations of the underlying network.

4.2.2   INFERENCE

We have reviewed methods which can be used to construct graphs over large datasets, and how such graphs can be partitioned to improve inference speeds in SMP and distributed computer settings. In this section, we turn our attention to how label inference may be efficiently performed over such large graphs.

MapReduce-based Inference: As mentioned in Chapter 3, many graph-based SSL algorithms, e.g., MP, IR, GRF, LGC, Adsorption, and MAD, are iterative in nature where labels on a given node are updated based on the current label distribution of the immediate neighborhood of the given node. Even though the individual update of each algorithm varies, access to the label information in the immediate neighborhood, in addition to the information on the given node, is all that is necessary. This makes such algorithms easily parallelizable, especially in the MapReduce framework [Dean and Ghemawat, 2008] as follows.

- **Map Phase**: In the MAP phase, each node sends its current label distribution to all its immediate neighbors. Please note that the neighbor of a given node may be partitioned to be in the same computer as the source node, or in some other computer accessible over the network.

- **Reduce Phrase**: In the Reduce phase, each node updates its label information based on the messages it received from its neighbors, its own seed labels, and any regularization targets. Completion of all the nodes in the graph completes one iteration of the graph SSL algorithm, and the Map-Reduce phases are repeated until convergence.

Apache Hadoop is an open-source implementation of the MapReduce framework. Thus, the algorithms mentioned above may be implemented in Hadoop. Hadoop, and MapReduce in general, has been shown to scale to large datasets. Thereby, such parallel implementation of graph SSL algorithms can scale to very large datasets and graphs. In fact, the Junto Label Propagation Toolkit [2] already provides standalone and Hadoop-based implementations of a few of these algorithms.

Other Methods: Karlen et al. [2008] describe a method to speed up Transductive Support Vector Machines (TSVM). Tsang and Kwok [2006] propose a sparse and scalable framework for manifold regularization.

4.3   SCALING TO LARGE NUMBER OF LABELS

Thus far we have considered how to scale graph-based SSL algorithms to problems with relatively large number of samples. In this section we focus on problems with a large number of *labels* (m). Majority of the graph-based SSL algorithms discussed in Chapter 3 scale linearly in the amount of space required to store the label distributions. Many practical applications involve a very large number of labels and in such situations it is important to pay attention to space and time complexity trade-offs. For instance, Carlson et al. [2010] describe an SSL system with hundreds of overlapping classes, and Shi et al. [2009] present a text classification task with over 7000 classes. The ImageNet dataset [Deng et al., 2009] poses a classification task involving 100,000 classes. In order to address these situations [Talukdar and Cohen, 2014] recently proposed MAD-SKETCH, a graph-based SSL method which can scale to large number of potentially overlapping labels. The core idea is to represent label scores at each node using *Count-min Sketch (CMS)* [Cormode and Muthukrishnan, 2005], a randomized data structure. We first provide a brief overview of CMS and then describe how MAD-SKETCH uses them to scale to large number of labels.

Count-min Sketch (CMS)   is a widely-used probabilistic scheme. It stores an approximate mapping between integers i and associated real values y_i. Specifically, the count-min sketch consists of a $w \times d$ matrix $\mathbb{S}$, together with d hash functions $h_1, \ldots, h_d$, which are chosen randomly from a pairwise-independent family. A sketch is always initialized to an all-zero matrix. Let $\mathbf{y}$ be a sparse vector, in which the i-th component is denoted by y_i. To store a new value y_i to be associated with component i, one simply updates as follows:

$$\mathbb{S}_{j,h_j(i)} \leftarrow \mathbb{S}_{j,h_j(i)} + y_i, \ \forall 1 \le j \le d.$$

To retrieve an approximate value $\hat{y}_i$ for i, one computes the minimum value over all $j : 1 \le j \le d$ of $\mathbb{S}_{j,h_j(i)}$ as follows:

$$\tilde{y}_i = \min_{1 \le j \le d} \mathbb{S}_{j,h_j(i)}. \tag{4.1}$$

We note that $\tilde{y}_i$ will never underestimate y_i, but may overestimate it if there are hash collisions. The parameters w and d can be chosen so that with high probability, the overestimation error is small. Note that count-min sketches are linear. In other words, if $\mathbb{S}_1$ is the matrix for the count-min sketch of $\mathbf{y}_1$ and $\mathbb{S}_2$ is the matrix for the count-min sketch of $\mathbf{y}_2$, then $a\mathbb{S}_1 + b\mathbb{S}_2$ is the matrix for the count-min sketch of the vector $a\mathbf{y}_1 + b\mathbf{y}_2$.

[2]The Junto Label Propagation Toolkit: https://github.com/parthatalukdar/junto

MAD-Sketch using CMS   From Chapter 3 we recall that Modified Adsorption (MAD) uses the following iterative update to estimate the labels scores at each node:

$$\hat{Y}_v^{(t+1)} \leftarrow \frac{1}{M_{vv}}(\mu_1 \times S_{vv} \times Y_v + \mu_2 \times D_v^{(t)} + \mu_3 \times R_v), \ \forall v \in V$$
$$\text{where } D_v^{(t)} = \sum_{u \in V}(W_{uv}' + W_{vu}') \times \hat{Y}_u^{(t)},$$
$$\text{and } M_{vv} = \mu_1 S_{vv} + \mu_2 \sum_{u \neq v}(W_{uv}' + W_{vu}') + \mu_3. \tag{4.2}$$

We will refer to this version as MAD-EXACT since the labels and their scores are explicitly stored in each node. Instead of storing label scores explicitly, MAD-SKETCH uses Count-min Sketch (CMS) to compactly store label scores at nodes. In particular, MAD-SKETCH uses the following iterative update to estimate label scores at each node, with iterations run until convergence.

$$\mathbb{S}_{\hat{Y},v}^{(t+1)} \leftarrow \frac{1}{M_{vv}}\left(\mu_1 \times S_{vv} \times \mathbb{S}_{Y,v} + \mu_2 \times \sum_{u \in V}(W_{uv}' + W_{vu}') \times \mathbb{S}_{\hat{Y},u}^{(t)} + \mu_3 \times \mathbb{S}_{R,v}\right), \ \forall v \in V \tag{4.3}$$

where $\mathbb{S}_{\hat{Y},v}^{(t+1)}$ is the count-min sketch corresponding to label score estimates on node v at time $t + 1$, $\mathbb{S}_{Y,v}$ is the sketch corresponding to any seed label information, and finally $\mathbb{S}_{R,v}$ is the sketch corresponding to label regularization targets in node v. Please note that due to linearity of CMS, we do not need to unpack the labels for each update operation. This results in significant run-time improvements compared to non-sketch-based MAD-EXACT [Talukdar and Cohen, 2014]. After convergence, MAD-SKETCH returns $\mathbb{S}_{Y,v}$ as the count-min sketch containing estimated label scores on node v. The final label score of a label can be obtained by querying this sketch using Equation 4.1. Let $\tilde{Y}$ be this matrix of estimated labels scores derived from the sketch.

   We now turn to the question of how well MAD-SKETCH approximates the exact version—i.e., how well $\tilde{Y}$ approximates $\hat{Y}$. We begin with a basic result on the accuracy of count-min sketches [Cormode and Muthukrishnan, 2005].

Theorem 4.1 [Cormode and Muthukrishnan, 2005].   *Let y be an vector, and let $\tilde{y}_i$ be the estimate given by a count-min sketch of width w and depth d for y_i. If $w \geq \frac{e}{\eta}$ and $d \geq \ln(\frac{1}{\delta})$, then $\tilde{y}_i \leq y_i + \eta\|y\|_1$ with probability at least 1-δ.*

   To apply this result, we consider several assumptions that might be placed on the learning task. One natural assumption is that the initial seed labeling is k-sparse, which we define to be a labeling so that for all v and ℓ, $\|Y_v.\|_1 \leq k$ for some small k (where $Y_v.$ is the v-th row of Y). If $\tilde{Y}$ is a count-min approximation of $\hat{Y}$, then we define the approximation error of $\tilde{Y}$ relative to $\hat{Y}$ as

Table 4.1: Table comparing average per-iteration memory usage (in GB), runtime (in seconds), and Mean Reciprocal Rank (MRR) of MAD-Exact and MAD-Sketch (for various sketch sizes) when applied over a dataset derived from Freebase. Using sketch size of $w = 109$ and $d = 8$ as prescribed by Theorem 4.2 for this dataset, we find that MAD-Sketch is able to obtain the same MRR (performance) as MAD-Exact, while using a reduced memory footprint and achieving about 4.7 speedup. Additionally, we observe that even more aggressive sketches (e.g., $w = 20, d = 8$ in this case) may be used in practice, with the possibility of achieving further gains in memory usage and runtime. Table is reproduces from Talukdar and Cohen [2014]

	Average Memory Usage (GB)	Total Runtime (s) [Speedup w.r.t. MAD-Exact]	MRR
MAD-Exact	3.54	516.63 [1.0]	0.28
MAD-Sketch ($w = 109, d = 8$)	2.68	110.42 [4.7]	0.28
MAD-Sketch ($w = 109, d = 3$)	1.37	54.45 [9.5]	0.29
MAD-Sketch ($w = 20, d = 8$)	1.06	47.72 [10.8]	0.28
MAD-Sketch ($w = 20, d = 3$)	1.12	48.03 [10.8]	0.23

$\max_{v,\ell}(\tilde{Y}_{v,\ell} - \hat{Y}_{v,\ell})$. Note that approximation error cannot be negative, since a count-min sketch can only overestimate a quantity.

Theorem 4.2 [Talukdar and Cohen, 2014]. *If the parameters of MAD-Sketch μ_1, μ_2, μ_3 are such that $\mu_1 + \mu_2 + \mu_3 \leq 1$, Y is k-sparse and binary, and sketches are of size $w \geq \frac{ek}{\epsilon}$ and $d \geq \ln\frac{m}{\delta}$ then the approximation error MAD-Sketch is less than ϵ with probability $1-\delta$.*

Although this result is stated for MAD-Sketch, it also holds for other label propagation algorithms that update scores by a weighted average of their neighbor's scores, such as the harmonic method [Zhu et al., 2003]. Additional results in the settings with skewed label scores and community graph structure are also presented in Talukdar and Cohen [2014].

Experimental results comparing MAD-Exact and MAD-Sketch are reproduced from Talukdar and Cohen [2014] in Table 4.1 and Figure 4.3. These results show that the use of sketches in MAD-Exact not only reduces space requirements, but also improves runtime. Please see Section 6 of Talukdar and Cohen [2014] for additional experiments and analysis.

4.4   CONCLUSIONS

In this chapter, we have presented various approaches for scaling graph-based SSL algorithms to large datasets. While k-NN is the most popular approach for graph construction, brute-force k-NN is quadratic in the number of samples and thus does not scale well to large problems. Approximate nearest neighbor search has been the topic of extensive research [Bentley, 1975], with many algorithms able to achieve $O(\log n)$ or faster performance [Arya and Mount, 1993,

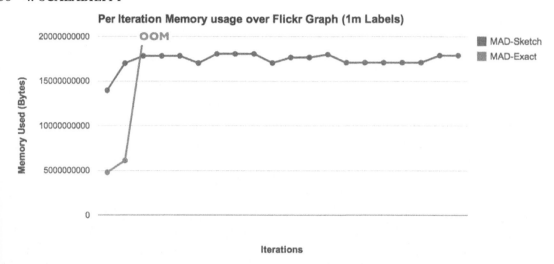

Figure 4.3: Per-iteration memory usage by MAD when labels and their scores on each node are stored exactly (MAD-Exact) vs. using Count-Min Sketch (MAD-Sketch) in the Flickr-1m dataset, with total unique labels $m = 1,000,000$, and sketch parameters $w = 55$, $d = 17$. We observe that even though MAD-Exact starts out with a lower memory footprint, it runs out of memory (OOM) by third iteration, while MAD-Sketch is able to compactly store all the labels and their scores with almost constant memory usage over the 20 iterations. Figure is reproduced from Talukdar and Cohen [2014].

Arya et al., 1998, Friedman et al., 1977]. Despite decades of research, the current nearest neighbor search algorithms are exponential in the number of input dimensions in either space or time. In practice, as the number of dimensions increases, approximate nearest neighbor search algorithms provide little or in some case no improvement over brute-force search. Recently, however, locality sensitive hashing (LSH) has been shown to scale well for high-dimensional inputs [Andoni and Indyk, 2008].

Yet another way to scale graph construction is to consider only a subset of the samples. This not only speeds up graph construction but also has the nice side-effect of speeding up inference. However, choosing (sampling) the subset is challenging and most of the proposed approaches use heuristics to generate the samples. Sampling is further complicated in the case of sparse high-dimensional features.

We also presented approaches to scale inference in graph-based SSL algorithms. Inference in the case of a majority of the graph-based SSL algorithms is iterative and can be expressed as simple and efficient messages passed along the edges of the graph. As a result the manner in which the graph is organized plays a critical role. We discussed graph ordering (or partitioning) approaches in a single machine with multiple threads, i.e., a shared-memory symmetric multi-processor (SMP), and also in the case of a distributed computing environment. The results show that careful graph ordering where there is a high-degree of neighbor overlap can lead to linear speedups in the SMP case. The distributed computing environment presents some challenges as inter-process communication is very expensive and therefore one needs to organize the nodes in the graph so that such communication is as small as possible. We also showed how to scale-up some of graph-based SSL algorithms using MapReduce. This takes advantage of the fact that the update for a node in many graph-based SSL algorithms is a function of the scores of its neighbors.

Finally we presented an approach to scale graph-based SSL algorithms to problems that have a very large number of output labels. This involves representing the label scores at each node using a randomized data structure, count-min sketch [Cormode and Muthukrishnan, 2005]. While there has been a lot of progress in the area of scaling up graph-based SSL algorithms, there are still quite a few open question in the area. This will be discussed in Chapter 6.

CHAPTER 5

Applications

In this chapter we describe some of the applications in which graph-based approaches have been used. Our descriptions here are not intended to be exhaustive enough to duplicate all the experiments and results. Rather, the goal here is to present the different applications and highlight salient aspects. In each case we briefly describe the task, data set, how the graph is constructed, methods used and results. For more details, readers should look at the original publications.

5.1 TEXT CLASSIFICATION

Given a document (e.g., web-page, news article), the goal of text classification is automatically to assign the document to a fixed number of semantic categories. Each document may belong to one, many, or none of the categories. In general, text classification is a *multi-class* problem. Training fully supervised text classifiers requires large amounts of labeled data whose annotation can be expensive [Dumais et al., 1998]. As a result there has been interest is using SSL techniques for text classification [Joachims, 1999, 2003, Orbach and Crammer, 2012, Subramanya and Bilmes, 2008, Talukdar and Crammer, 2009].

Data

Reuters-21578   The standard setup is to use "ModApte" split of the Reuters-21578 data set collected from the Reuters newswire in 1987 [Lewis et al., 1987]. The corpus has 9,603 training and 3,299 test documents. Of the 135 potential topic categories only the 10 most frequent categories are used [Joachims, 1999]. Categories outside the 10 most frequent are collapsed into a "other" class.

WebKB   World Wide Knowledge Base (WebKB) is a collection of 8282 web pages obtained from four academic domains. The web pages in the WebKB set are labeled using two different polychotomies. The first is according to topic and the second is based on web domain. In the experiments described below, only the first polychotomy is considered, which consists of 7 categories: *course*, *department*, *faculty*, *project*, *staff*, *student*, and *other*. Following Nigam et al., Subramanya and Bilmes [2008] only use documents from categories *course*, *department*, *faculty*, *project* which gives 4199 documents for the four categories. Each of the documents is in HTML format containing text as well as other information such as HTML tags, links, etc.

Table 5.1: Precision Recall Break Even Points (PRBEP) for the WebKB data set with $n_l = 48$ and $n_u = 3148$. The last row is the macro-average over all the classes. In each row the best result is shown in bold. Table is reproduced from Talukdar and Crammer [2009] and Subramanya and Bilmes [2008]

Class	SVM	TSVM	SGT	GRF	MP	Adsorption	MAD
course	46.5	43.9	29.9	45.0	**67.6**	61.1	67.5
faculty	14.5	31.2	42.9	40.3	42.5	**52.8**	42.2
project	15.8	17.2	17.5	27.8	42.3	**52.6**	45.5
student	15.0	24.5	56.6	51.8	55.0	39.8	**59.6**
average	23.0	29.2	36.8	41.2	51.9	51.6	**53.7**

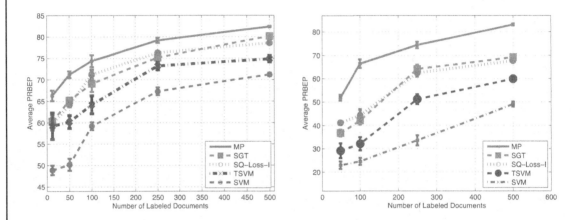

Figure 5.1: Average PRBEP over all classes vs. number of labeled documents (n_l) for the Reuters (left) and WebKB (right) data sets. Figure is reproduced from Subramanya and Bilmes [2008].

Feature Extraction & Graph Construction

All stop-words (e.g., "the," "that") and case information are removed from each document $i \in \mathcal{D}$. Next inflection information is removed (e.g., "fishing" becomes "fish"); this commonly referred to as *stemming* Porter [1997]. Each document, i, is represented by a bag-of-words feature vector $\mathbf{x}_i = (x_1, \ldots, x_d)$ where d is the size of the vocabulary. x_k is set to the TFIDF score [Salton and Buckley, 1987] of the word in the vocabulary (indexed by k).

Each node in the graph represents a document encoded by the feature vector $\mathbf{x}_i$. Thus, the goal of the graph construction process is to find similar documents. Symmetrized k-NN graphs are constructed with the weights generated using cosine similarity.

Results

In text categorization, precision-recall break even point (PRBEP) is used to measure the performance of an algorithm. It is defined as that value for which precision and recall are equal.

Comparison of the results on the WebKB data set are shown in Table 5.1. Methods include SVM [Joachims, 1999], transductive SVM (TSVM) [Joachims, 1999], spectral graph transduction (SGT) [Joachims, 2003], GRF [Zhu et al., 2003], measure propagation (MP) [Subramanya and Bilmes, 2008], and Adsorption and Modified Adsorption (MAD) [Talukdar and Crammer, 2009]. Note that SVM is a fully supervised method, TSVM is a semi-supervised learning method but does not rely on a graph, while all other remaining approaches fall under the graph-based SSL algorithm category. Furthermore, as TSVM and SGT are binary classification approaches and this is a multi-class problem, they are implemented in a one vs. rest manner [Joachims, 2003]. While it is possible to train multi-class SVMs [Crammer and Singer, 2002], the results in Table 5.1 are for a binary SVM, again implemented in a one vs. rest manner.

It can be seen that all the SSL approaches outperform the supervised SVM. This is not surprising as the supervised approach is unable to take advantage of the unlabeled data. Further, the graph-based SSL approaches perform better than TSVM which does not make use of a graph. Amongst the graph-based approaches, MAD has best macro-average PRBEP. As there are 4 classes and $n_l = 48$, it implies about 16 labeled documents per class. It can be seen that one can achieve reasonable performance by annotating only 16 documents per class by taking advantage of the unlabeled data.

The macro-averaged PRBEP vs. the number of labeled documents is shown in Figure 5.1. The plot on the left is for the Reuters data set while the plot on the right shows the results for WebKB. It can be seen that the performance of the different approaches increases with more labeled data. However, the SSL approaches show bigger gains compared to the supervised SVM. Further, graph-based approaches perform the best.

5.2   PHONE CLASSIFICATION

A *phone* is a basic unit of speech. Given an input frame (or segment) of speech, the goal of phone classification is to categorize it into one among a fixed set of classes [Halberstadt and Glass, 1997]. As a phone is the basic building block of speech, phone classification is an important step towards automatic speech recognition. Further, physiological constraints on the speech articulators limit the degrees of freedom of the organs involved in speech production. As a result, humans are only capable of producing sounds occupying a subspace of the entire acoustic space [Errity and McKenna, 2006]. Thus, speech data can be viewed as lying on a low dimensional manifold embedded in the original acoustic space thus making speech more suitable for graph-based learning approaches.

Data
TIMIT [Zue et al., 1990] is a corpus of read speech designed to provide speech data for acoustic-phonetic studies. It contains recordings of 630 speakers from 8 major dialects of American English, each reading 10 phonetically rich sentences.

Switchboard I   (SWB) [Godfrey et al., 1992] is a collection of 2,400 telephone conversations among 543 speakers from all areas of the United States. It consists of about 300 hours of speech. The *Switchboard Transcription Project* (STP) [Greenberg, 1995] annotated a small part of the SWB corpus at the phone and syllable levels. As the task is time-consuming, costly, and error-prone, only 75 min of speech segments were annotated at the phone level.

Feature Extraction and Graph Construction

TIMIT: To obtain $\mathbf{x}_i$, the speech signal is first pre-emphasized and then windowed using a Hamming window of size 25 ms at 100 Hz. One then extracts mel-frequency cepstral coefficients (MFCCs) [Lee and Hon, 1989] from these windowed features. As speech is a continuous time signal, phone classification performance is improved by using more contextual information and so features from the immediate left and right contexts are appended.

The above features were used to construct a symmetrized k-NN graph over the standard TIMIT training and development sets. The weights are given by

$$w_{ij} = k(\mathbf{x}_i, \mathbf{x}_j) = \exp\{-(\mathbf{x}_i - \mathbf{x}_j)^\top \Sigma^{-1}(\mathbf{x}_i - \mathbf{x}_j)\},$$

where Σ is the covariance matrix and is given by

$$\Sigma = \frac{1}{n}\sum_{i=1}^{n}(\mathbf{x}_i - \mu)^\top(\mathbf{x}_i - \mu) \text{ where } \mu = \frac{1}{n}\sum_{i=1}^{n}\mathbf{x}_i.$$

SWB: Here, features were extracted in the following manner—the input wave files were first segmented and then windowed using a Hamming window of size 25 ms at 100 Hz. Subsequently, 13 perceptual linear prediction (PLP) coefficients from these windowed features. As in the case of TIMIT, contextual information is appended to the features as it leads to better phone classification performance.

In the case of SWB, there are about 120 million input instances, i.e., $n = 120{,}000{,}000$. As a result this corpus is not amenable to "brute-force" graph construction methods which are computationally $O(n^2)$. Thus, Subramanya and Bilmes [2010] construct the graph using kd-tree based data structures [Friedman et al., 1977]. They make use of the Approximate Nearest Neighbor (ANN) library [Arya and Mount, 1993, Arya et al., 1998].

Results

In the case of TIMIT, the graph is constructed over the standard TIMIT training and development sets. The graph had about 1.4 million vertices. Subramanya and Bilmes [2010] simulate the semi-supervised setting by treating fractions of the TIMIT training set as labeled. Phone classification results are shown in Figure 5.2. The plot on the left shows the results on the TIMIT development set. This simulates the case when only small amounts of data are labeled. The y-axis shows phone accuracy (PA) which represents the percentage of frames correctly classified and

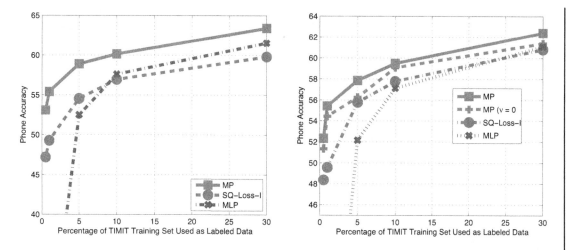

Figure 5.2: Phone accuracy (PA) on the TIMIT development set (left) and TIMIT NIST core evaluation/test set (right). The x-axis shows the percentage of standard TIMIT training data that was treated as being labeled. Figures are reproduced from Subramanya and Bilmes [2010].

the x-axis shows the fraction of the training set that was treated as being labeled. They compare the performance of measure propagation (MP) against GRF and a fully supervised 2-layer multi-layered perceptron (MLP) [Bishop, 1995]. It can be seen that both the graph-based SSL approaches outperform the supervised MLP. The difference between the performance of the supervised MLP and semi-supervised approaches is larger in conditions where there are very small amounts of labeled data. Thus, either in the case of new domains or new languages where only limited amounts of labeled data are available, semi-supervised and in particular graph-based approaches are likely to lead to better performance compared to fully supervised approaches. Further, the results also show that MP outperforms GRF.

While MP and GRF are both transductive, Subramanya and Bilmes [2010] show how to extend them unseen test data, i.e., inductive extensions. Given a test sample $\hat{\mathbf{x}} \notin \mathcal{D}$, they make use of the Nadaraya-Watson estimator to estimate the output label using

$$\hat{y} = \underset{y \in \mathcal{Y}}{\operatorname{argmax}}\, \hat{p}(y) \text{ where } \hat{p}(y) = \frac{\sum_{j \in \Gamma(\hat{\mathbf{x}})} k(\hat{\mathbf{x}}, \mathbf{x}_j) p_j^*(y)}{\sum_{j \in \Gamma(\hat{\mathbf{x}})} k(\hat{\mathbf{x}}, \mathbf{x}_j)},$$

where $\Gamma(\hat{\mathbf{x}})$ are the nearest neighbors of $\hat{\mathbf{x}}$ in $\mathcal{D}$. The results of the inductive extension on the TIMIT test set are shown in Figure 5.2 (right plot). Similar to the transductive case it can be seen that the graph-based SSL approaches outperform the fully supervised approaches.

Kirchhoff and Alexandrescu [2011] present results of using modified adsorption on the TIMIT corpus. However, they take a slightly different approach to construct the graph—they first train a fully supervised classifier (MLP) on the available labeled data. The graph is then

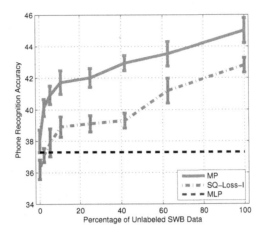

Figure 5.3: Phone accuracy vs. percentage of switchboard (SWB) I training data. STP portion of SWB was excluded. Phone accuracy was measured on the STP data. Note that when all the Switchboard I data was added, the resulting graph had **120 million** vertices. The dashed black line shows the performance of a MLP measured using the $s = 0\%$ case over the same training, development and test sets as MP and LP. Figure is reproduced from Subramanya and Bilmes [2010].

constructed by using a kernel defined over the posteriors output by the trained classifier. They argue that the classifier removes noise in the data and results in a new feature space where classes are clustering more tightly leading to better graphs [Alexandrescu and Kirchhoff, 2007]. Also unlike the case of Subramanya and Bilmes [2010] they construct the graph over the training, development and test sets. The results show that MAD outperforms the GRF objective by about 1% when using 10% of the labeled instances in the TIMIT training set.

Subramanya and Bilmes [2010] investigate the effect of adding unlabeled data to SSL performance using the SWB corpus. They split the 75 min phonetically annotated part of STP into a training, development and test set. The development set is used to tune the relevant hyperparameters. The results on the test set are shown in Figure 5.3. The y-axis represents the phone recognition accuracy while the x-axis shows the percentage of *unlabeled* data. The compare the performance of a fully supervised MLP against the graph-based MP and GRF algorithms. As expected, the performance of the MLP does not change with the amount of the unlabeled data. However, the SSL algorithms are able to take advantage of the unlabeled data and their performance improves as more unlabeled data is used. This is counter to the theoretical analysis of Nadler et al. [2010] showing that under conditions of infinite amounts of unlabeled data, the GRF objective converges to the uniform distribution. Perhaps the amount of unlabeled data used has not reached the point where it begins to be harmful. Another interesting aspect of this is that for the setting where they use all the unlabeled data which leads to a graph with about 120 million vertices, thus showing that graph-based algorithms can easily be scaled to very large problems.

5.3   PART-OF-SPEECH TAGGING

Given a sentence, the process of inferring the part-of-speech (e.g., noun, verb, adjective, etc.) for each word in the sentence is referred to as part-of-speech (POS) tagging. The POS tag of a word depends both on the word and its context. Further, there are constraints on the sequence of POS tags for the words in a sentence. Thus, POS tagging is a structured prediction problem, in that the output space has an inherent structure. A large number of problems in speech and natural language processing such as speech recognition, named-entity recognition, machine translation, and syntactic parsing have a structured output space. Annotating data for these tasks is cumbersome and often error-prone making them ideal candidates for SSL.

All the SSL applications that we have discussed thus far have focused on unstructured classification problems, that is, problems with a relatively small set of atomic labels. In the case of POS tagging, while the tag for each word comes from a relatively small number of POS tags, there are exponentially many ways to combine them into a final structured label. The use of graph-based methods to solve a structured SSL problem presents some challenges including, representation, graph construction and inference.

Recall that we represent labeled data by $\mathcal{D}_l = \{(\mathbf{x}_i, y_i)\}_{i=1}^{n_l}$, and unlabeled data is given by $\mathcal{D}_u = \{\mathbf{x}_i\}_{i=n_l+1}^{n_l+n_u}$. In this application, $\mathbf{x}_i$ represents a sequence of words given by $\mathbf{x}_i = x_i^{(1)} x_i^{(2)} \cdots x_i^{(|\mathbf{x}_i|)}$. As a result the corresponding output y_i is a sequence of POS tags given by $y_i = y_i^{(1)} y_i^{(2)} \cdots y_i^{(|\mathbf{x}_i|)}$ where $y_i^{(j)} \in \mathcal{Y}$, $\mathcal{Y}$ is the set of POS tags and thus $y_i \in \mathcal{Y}^{|\mathbf{x}_i|}$. Both $\mathbf{x}_i$ and y_i are sequences of length $|\mathbf{x}_i|$.

Linear-chain Conditional Random Fields   (CRF) [Lafferty et al., 2001, Sutton and Mccallum, 2006] model the output sequence given an input in the following manner:

$$p(y_i|\mathbf{x}_i; \Lambda) \propto \exp \left(\sum_{j=1}^{|\mathbf{x}_i|} \sum_{k=1}^{K} \lambda_k f_k(y_i^{(j-1)}, y_i^{(j)}, \mathbf{x}_i, j) \right) \quad .$$

Here, $\Lambda = \{\lambda_1, \ldots, \lambda_K\} \in \mathcal{R}^K$, $f_k(y_i^{(j-1)}, y_i^{(j)}, \mathbf{x}_i, j)$ is the k-th feature function applied to two consecutive CRF states, $y_i^{(j-1)}$ and $y_i^{(j)}$, and the input sequence; λ_k is the weight of that feature. Given some labeled data the optimal feature weights are given by:

$$\Lambda^* = \operatorname*{argmin}_{\Lambda \in \mathbb{R}^K} \left[-\sum_{i=1}^{n_l} \log p(y_i|\mathbf{x}_i; \Lambda) + \gamma \sum_{k=1}^{K} \lambda_k^2 \right] \quad . \tag{5.1}$$

The above objective can be optimized using standard convex optimization algorithms such as L-BFGS [Wallach, 2002]. The above objective is fully supervised. CRFs have been used in a number of applications in speech and language [McCallum and Li, 2003, Sung et al., 2007].

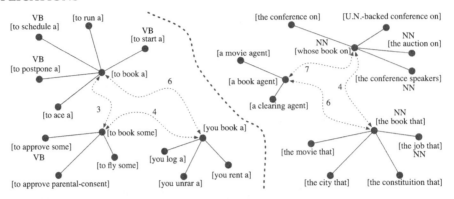

Figure 5.4: Vertices with center word "book" and their local neighborhoods, as well as the shortest-path distance between them. Note that the noun (NN) and verb (VB) interpretations form two disjoint connected components. Figure is reproduced from Subramanya et al. [2010].

Data

Subramanya et al. [2010] use the Wall Street Journal (WSJ) section of the Penn Treebank as labeled data. The WSJ data comes from the newswire domain. While POS taggers achieve accuracies similar to the level of inter-annotator agreement (> 97%) on in-domain test sets [Toutanova et al., 2003], performance on out-of-domain data is often well below 90%. While this may seem good enough, POS tagging is usually the first step in many NLP pipelines and thus POS tagging errors are highly detrimental to overall performance. As a result, adapting taggers trained on data from the newswire domain to other target domains is an important task. Subramanya et al. [2010] have proposed an algorithm to adapt a POS tagger using SSL techniques. Here the unlabeled data comes from the target domain and the goal is to modify the parameters of the model to better suit the target domain. They consider two target domains: questions and biomedical data.

Question Bank   [Judge et al., 2006] is a corpus of 4,000 questions that are manually annotated with POS tags and parse trees. Questions are difficult to tag with WSJ-trained taggers primarily because the word order is very different compared to the mostly declarative sentences in the training data. The unlabeled data for this domain consists of a set of 100,000 questions collected from anonymized Internet search queries.

BioTreebank   is a corpus of 1,061 sentences with manually annotated POS tags. In general biomedical text presents a number of challenges to POS taggers trained on WSJ data. The unlabeled data are 100,000 sentences that were chosen by searching MEDLINE for abstracts pertaining to cancer, in particular genomic variations and mutations [Blitzer et al., 2006].

Table 5.2: Features extracted for the word sequence "x_1 x_2 x_3 x_4 x_5" where the trigram is "x_2 x_3 x_4." Table reproduced from Subramanya et al. [2010]

Description	Feature
Trigram + Context	x_1 x_2 x_3 x_4 x_5
Trigram	x_2 x_3 x_4
Left Context	x_1 x_2
Right Context	x_4 x_5
Center Word	x_2
Trigram—Center Word	x_2 x_4
Left Word + Right Context	x_2 x_4 x_5
Left Context + Right Word	x_1 x_2 x_4
Suffix	HasSuffix(x_3)

Feature Extraction & Graph Construction

Graph construction in problems where the input has an underlying structure poses a number of challenges. Firstly, there is the question of how to represent the input using a graph. In POS tagging, each node in the graph could represent a word in the sentence or the entire sentence. While it is possible to use an appropriate similarity metric over complete sequences, such as edit distance or a string kernel, to construct the graph in cases where the entire sentence is represented by a node, it is not clear how to use the complete input sequence similarity to constrain the output tag sequence. On the other hand, choosing a representation in which each node in the graph represents a single word in the input sequence implies that all contextual information is lost thus making it hard to learn from the unlabeled data.

Altun et al. [2005] were the first to propose a scheme for constructing graphs for structured SSL problems by considering *parts* of structured examples (also known as factors in graphical model terminology), which encode the local dependencies between input data and output labels in the structured problem. Subramanya et al. [2010] also propose a similar approach by making use of local contexts as graph vertices. Their approach exploits the empirical observation that the POS of a word is mostly determined by its local context. Specifically, the set V of graph vertices consists of all the word n-grams (*types*) that have occurrences (*tokens*) in both the labeled and unlabeled sentences. Thus, each vertex in the graph represents a n-gram type. In other words, while an n-gram may occur multiple times in the input corpus, it is represented by a single node in the graph. Subramanya et al. [2010] construct k-NN graphs over $n = 3$ or trigram types using cosine similarity between the feature vectors for the vertices. The features are described in Table 5.2.

Figure 5.4 shows an excerpt of the graph from Subramanya et al. [2010]. It is interesting to note that the neighborhoods are coherent, showing very similar syntactic configurations. Furthermore, vertices that should have the same label are close to each other, forming connected components for each part-of-speech category (for nouns and verbs in the figure).

Algorithm 6: Semi-Supervised CRF Training

$\Lambda_l^* = \text{crf-train}(\mathcal{D}_l, \Lambda^0)$
Set $\Lambda_0 = \Lambda_l^*$
while $t = 0, \cdots, T$ **do**
$\quad \{\mathbf{P}\} = \text{posterior\_decode}(\mathcal{D}_u, \Lambda_t)$
$\quad \{\mathbf{Q}\} = \text{token\_to\_type}(\{\mathbf{P}\})$
$\quad \{\hat{\mathbf{Q}}\} = \text{graph\_propagate}(\{\mathbf{Q}\})$
$\quad \mathcal{D}_u^{(t)} = \text{viterbi\_decode}(\{\hat{\mathbf{Q}}\}, \Lambda_t)$
$\quad \Lambda_{t+1} = \text{crf-train}(\mathcal{D}_l \cup \mathcal{D}_u^{(t)}, \Lambda_t)$
end while
Set $\Lambda_{l \cup u}^* = \Lambda_T$
Return last $\Lambda_{l \cup u}^*$

Results

Subramanya et al. [2010] propose Algorithm 6 to train CRFs in a semi-supervised manner using graph-based methods. They first train a CRF using the labeled data whose parameters, denoted by Λ_l^*, are used to initialize the semi-supervised training. The iterative SSL algorithm consists of five steps which we describe briefly. In *posterior decoding*, the current estimates of the parameters of the CRF are used to compute the marginal probabilities $p(y_i^{(j)}|\mathbf{x}_i; \Lambda_n^{(t)})$ over POS tags for every word j in every sentence i in $\mathcal{D}_l \cup \mathcal{D}_u$.

As the graph is over trigrams, $p(y_i^{(j)}|\mathbf{x}_i; \Lambda_n^{(t)})$ is used as the posterior for the trigram sequence $x_i^{(j-1)}x_i^{(j)}x_i^{(j+1)}$. However, this trigram may occur multiple times in $\mathcal{D}$. The posteriors for each of these instances are aggregated in the *token to type* mapping step. Recall that each vertex in the graph represents a trigram type. For a sentence i and word position j in that sentence, let $T(i, j) \in V$ denote the trigram centered at position j. Conversely, for a trigram type u, let $T^{-1}(u)$ return all pairs (i, j) where i is the index of a sentence where u occurs and j is the position of the center word of an occurrence of u in that sentence. The type-level posteriors are computed using:

$$q_u(y) \triangleq \frac{1}{|T^{-1}(u)|} \sum_{(i,j) \in T^{-1}(u)} p(y_i^{(j)}|\mathbf{x}_i; \Lambda_n^{(t)}) \quad .$$

Given the graph and the marginals for each the vertices computed above, the graph propagation stage minimizes the following convex objective:

$$\mathbf{C}(q) = \sum_{u \in V_l} \|r_u - q_u\|^2 + \mu \sum_{u \in V, v \in \Gamma(u)} w_{uv}\|q_u - q_v\|^2 + \nu \sum_{u \in V} \|q_u - U\|^2$$
$$\text{s.t.} \sum_y q_u(y) = 1 \ \forall u \ \& \ q_u(y) \geq 0 \ \forall u, y,$$

where $\Gamma(u)$ is the set of neighbors of node u, and r_u is the empirical marginal label distribution for trigram u in the labeled data. The above objective is similar to the quadratic cost criterion [Bengio et al., 2007] and a special case of MAD [Talukdar and Crammer, 2009]. While the first term requires that the solution respect the labeled data, the second term enforces smoothness with respect to the graph while the last term acts as a regularizer. If an unlabeled vertex does not have a path to any labeled vertex, the last term ensures that the converged marginal for this vertex will be uniform over all tags.

Given the type marginals computed above, the next step interpolates them with the original CRF token marginals. This interpolation between type and token marginals encourages similar n-grams to have similar posteriors, while still allowing n-grams in different sentences to differ in their posteriors. For each unlabeled sentence i and word position j in it, the interpolated tag marginal is given by

$$\hat{p}(y_i^{(j)} = y | \mathbf{x}_i) = \alpha \times p(y_i^{(j)} = y | \mathbf{x}_i; \Lambda_t) + (1 - \alpha) \times q^*_{T(i,j)}(y), \qquad (5.2)$$

where α is a mixing coefficient which reflects the relative confidence between the original posteriors from the CRF and the smoothed posteriors from the graph. The interpolated marginals summarize all the information about the tag distribution for each word. However, using them on their own to select the most likely POS tag sequence, the first-order tag dependencies modeled by the CRF would be mostly ignored as the type marginals obtained from the graph after label propagation will have lost most of the sequence information. To enforce the first-order tag dependencies Viterbi decoding over the combined interpolated marginals and the CRF transition potentials is used to compute the best POS tag sequence for each unlabeled sentence. Let y_i^*, $i \in \mathcal{D}_u$ represent the 1-best sequence obtained after Viterbi decoding.

The last step of the iterative SSL involves using both the labeled data and the above inferred labels on the unlabeled data to re-train the supervised CRF using

$$\Lambda_{t+1} = \underset{\Lambda \in \mathbb{R}^K}{\operatorname{argmin}} \left[-\sum_{i=1}^{n_l} \log p(y_i | \mathbf{x}_i; \Lambda_t) - \eta \sum_{i=1}^{n_u} \log p(y_i^* | \mathbf{x}_i; \Lambda_t) + \gamma \|\Lambda\|^2 \right], \qquad (5.3)$$

where η and γ are hyper-parameters.

It is important to note that while each step of the above algorithm is convex, the combination is clearly not convex. Another interesting observation is that removing the graph propagation step in the above algorithm leads to a self-training like procedure. The graph, in addition to aiding learning using labeled data, also helps with providing information that cannot be expressed directly in a sequence model like a linear chain CRF. In particular, it is not possible in a CRF to directly enforce the constraint that similar trigrams appearing in different sentences should have similar POS tags. This is an important constraint during SSL.

Table 5.3 compares the results the fully supervised linear-chain CRF, semi-supervised CRF and self-trained CRF for both domains. For the question corpus, self-training improves over the baseline by about 0.6% on the development set. However, the gains from self-training are more

Table 5.3: Domain adaptation experiments. POS tagging accuracies in %. Table is reproduced from Subramanya et al. [2010]

	Questions		Bio	
	Dev	Eval	Dev	Eval
Supervised CRF	84.8	83.8	86.5	86.2
Self-trained CRF	85.4	84.0	87.5	87.1
Semi-supervised CRF	**87.6**	**86.8**	**87.5**	**87.6**

modest (0.2%) on the evaluation (test) set. The semi-supervised CRF is able to provide a more solid improvement of about 3% absolute over the supervised baseline and about 2% absolute over the self-trained system on the question development set. For the biomedical data, while the performances of semi-supervised CRF and self-training are statistically indistinguishable on the development set, we see modest gains of about 0.5% absolute on the evaluation set. On the same data, we see that the semi-supervised CRF provides about 1.4% absolute improvement over the supervised baseline. Subramanya et al. [2010] hypothesize that the lack of improvement on the Bio data set is primarily due to the sparse nature of the graph for that setting resulting for insufficient amounts of unlabeled data.

5.4   CLASS-INSTANCE ACQUISITION

Knowledge of classes and their instances, e.g., *angioplasty* (instance) is a *surgical procedure* (class), can be valuable in a variety of applications, ranging from web search to conference resolution. Traditionally, research within *supervised* named entity recognition has focused on a small number of coarse classes (viz., *person*, *location*, and *organization*). The need for heavy human supervision in the form of labeled data has been a bottleneck in scaling these techniques to large number of classes. Unfortunately, small number of classes is insufficient for many practical applications of interest. For example, in web search, users may be interested in symptoms of *malaria* or in the treatment of *cholera*, both of which are instances of class *diseases*. Such interest in a range of classes and their instances demand construction of datasets with large number of class-instance pairs.

Pattern-based information extraction systems [Hearst, 1992, Talukdar et al., 2006, Thelen and Riloff, 2002, Van Durme and Paşca, 2008] propose an interesting solution to this challenge. Starting with a few seed instances of a target class, these methods induce patterns from text which is turn are used to extract more instances of the same class from text. Next, some of the extracted instances are used to augment the seed set, and the process is repeated until some terminating condition is met. For example, given the seed instance *Pittsburgh* for class *location*, such methods might induce the pattern *"lives in __"*, which in turn could be used to extract additional instances such as *Chicago*, *London*, etc. Some subset of these new extractions may be used as seeds in the next round, and the process is repeated until no new extractions are possible, or if some terminating condition is met. However, output of such pattern-based approaches tend to be high-precision but low-coverage. This may be attributed to the fact that pattern-based methods usually rely on redundancy to estimate confidence, and in the face of uncertainty, high confidence thresholds result in lower recall. In this section, we shall explore how graph-based methods may be used to overcome these challenges to build large repositories of class-instance pairs.

Class-Instance Acquisition from Heterogeneous Sources

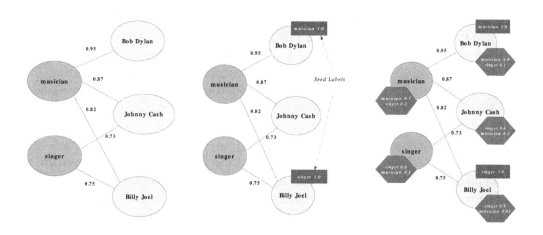

Figure 5.5: Section of a graph during various stages of the Adsorption label propagation algorithm as used in [Talukdar et al., 2008]. The initial graph without any labels on nodes is shown in the upper-left corner, while the final graph with labels estimated for all nodes using Adsorption is shown in lower-right corner. For better readability, instance nodes are shaded in yellow while class nodes are shaded in blue.

Talukdar et al. [2008] present a graph-based SSL method which combines extractions from multiple sources—natural language text [Van Durme and Paşca, 2008] and structured tables on the web [Cafarella et al., 2008]—to improve coverage of class-instance acquisition. This method combines class-instance extractions from different sources (and methods) into one bi-partite graph, with classes on one side and instances on the other. This graph-based representation is motivated by several factors.

- Graphs can represent complex relationships between classes and instances. For example, an ambiguous instance such as *Michael Jordan* could belong to the class of both *Professors* and *NBA players*. Similarly, an instance may belong to multiple nodes in the hierarchy of classes. For example, *Blue Whales* could belong to both classes *Vertebrates* and *Mammals*, because *Mammals* are a subset of *Vertebrates*.

- Extractions from multiple sources, such as Web queries, Web tables, and text patterns can be represented in a single graph.

- Graphs make explicit the potential paths of information propagation that are implicit in the more common local heuristics used for weakly-supervised information extraction. For example, if we know that the instance *Bill Clinton* belongs to both classes *President* and *Politician* then this should be treated as evidence that the class of *President* and *Politician* are related.

Each instance-class pair, denoted by (i, C), extracted using baseline methods (e.g., Van Durme and Paşca [2008] or Cafarella et al. [2008]) as described above, is represented as a weighted edge in the graph $G = (V, E, W)$, where V is the set of nodes, E is the set of edges and $W : E \to \mathbb{R}^+$ is the weight function which assigns a positive weight to each edge. In particular, for each (i, C, w) triple obtained from the set of base extractions, i and C are added to set V and (i, C) is added to E, with $W(i, C) = w$. The weight w represents the total score of all extractions with that instance and class. Sample of this graph is shown in upper-left-corner of Figure 5.5. Please note that this graph construction process is different from the approaches outlined in Chapter 2. In this graph representation, all nodes are treated in the same way, regardless of whether they represent instances or classes. In particular, all nodes can be assigned class labels. For an instance node, that means that the instance is hypothesized to belong to the class; for a class node, that means that the node's class is hypothesized to be semantically similar to the label's class.

Given a set of seed class-instance pairs, the corresponding nodes are injected with their seed classes (upper-right corner of Figure 5.5). Adsorption label propagation algorithm is then run over this graph, assigning class labels to both class and instance nodes. However, there need not be any necessary alignment between a class node and any of the (class) labels. Various stages of this label propagation is shown in the bottom row of Figure 5.5.

Experiments

Talukdar et al. [2008] experimented with whether increased coverage at comparable quality is achievable when extractions from a high-precision low-coverage extractor such as [Van Durme and Paşca, 2008] is combined with candidate extractions from structured data sources such as WebTables [Cafarella et al., 2008] (for details on the specific extraction methods, please see Talukdar et al. [2008]). By combining extractions from these two sources and using the graph construction technique described above, Talukdar et al. [2008] construct a graph with 1.4 million nodes and 65 million edges.

In the first experiment, Adsorption was run over the graph constructed above with $m = 9081$ class labels. Precision of top 100 Adsorption extractions from 5 classes which were not present in Van Durme and Paşca [2008] is shown in Table 5.4. From this table, we indeed observe that Adsorption is able to find new class-instance pairs at fairly high precision (in 4 out of 5 cases) which were not part of the original high-precision extractions from Van Durme and Paşca [2008]. Examples of a few random extractions from these classes is shown in Table 5.5.

Table 5.4: Precision of top 100 Adsorption extractions (for five classes) which were **not** present in A8. Table reproduced from Talukdar et al. [2008]

Class	Precision at 100 (new extractions)
Book Publishers	87.36
Federal Agencies	29.89
NFL Players	94.95
Scientific Journals	90.82
Mammal Species	84.27

Table 5.5: Random examples of top ranked extractions (for three classes) found by Adsorption which were not present in Van Durme and Paşca [2008]. Table reproduced from Talukdar et al. [2008]

Seed Class	Sample of Top Ranked Instances Discovered by Adsorption
Book Publishers	Small Night Shade Books, House of Anansi Press, Highwater Books, Distributed Art Publishers, Copper Canyon Press
NFL Players	Tony Gonzales, Thabiti Davis, Taylor Stubblefield, Ron DIxon, Rodney Hannah
Scientific Journals	Journal of Physics, Nature Structural and Molecular Biology, Sciences Sociales et Santé, Kidney and Blood Pressure Research, American Journal of Physiology—Cell Physiology

As noted previously, in this experiments, Adsorption ignores the type of nodes (i.e., whether a node is a class or instance node), and assigns seed class labels to class nodes. Thus, in addition to finding new instances to existing classes (obtained through class assignment on instance nodes), it is also able to discover related classes (obtained through seed class assignment on class nodes), all at no extra cost. Some of the top ranked class labels assigned by Adsorption for the seed class labels in shown in Table 5.6. To the best of our knowledge, this is one of the first systems which can discover class-instance and class-class relationships simultaneously.

Next, we look at results comparing performance on ability to assign labels to a fixed set of instances. For this, Van Durme and Paşca [2008] considered instances from 38 WordNet classes, reusing the graph from above. Adsorption was run with varying number of seeds (1, 5, 10, 25) per class. Mean Reciprocal Rank (MRR) is use as the evaluation metric in this experiment:

$$MRR = \frac{1}{|Q|} \sum_{v \in Q} \frac{1}{r_v}, \tag{5.4}$$

where $Q \subseteq V$ is the set of test nodes, and r_v is the rank of the gold label among the labels assigned to node v. Higher MRR reflects better performance. Experimental results are presented in Figure 5.6. This results highlights Adsorption's, and graph-based SSL techniques', in general,

Table 5.6: Top class labels ranked by their similarity to a given seed class in Adsorption. Table reproduced from Talukdar et al. [2008]

Seed Class	Non-Seed Class Labels Discovered by Adsorption
Book Publishers	small presses, journal publishers, educational publishers, academic publishers, commercial publishers
Federal Agencies	public agencies, governmental agencies, modulation schemes, private sources, technical societies
NFL Players	sports figures, football greats, football players, backs, quarterbacks
Scientific Journals	prestigious journals, peer-reviewed journals, refereed journals, scholarly journals, academic journals
Mammal Species	marine mammal species, whale species, larger mammals, common animals, sea mammals

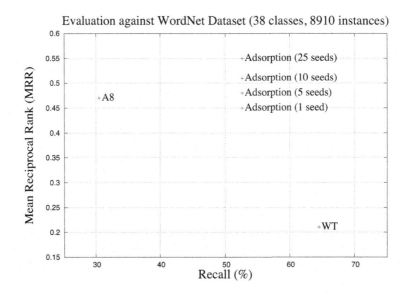

Figure 5.6: Plot of MRR vs. Recall comparing class-instance pairs extracted by [Van Durme and Paşca, 2008] from free text (referred to as A8 in the figure), from structured data [Cafarella et al., 2008], and Adsorption. Figure reproduced from Talukdar et al. [2008].

ability to effectively combine high-precision low-recall extractions with low-precision but high-recall extractions in a batter such that both precision and recall is improved, compared to the original high-quality extractions.

Comparative Study

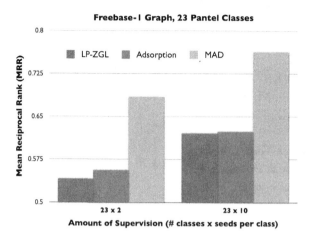

Figure 5.7:   Comparison of three graph transduction methods on a graph constructed from the Freebase dataset, with 23 classes. All results are averaged over four random trials. In each group, MAD is the rightmost bar. Figure is reproduced from Talukdar and Pereira [2010].

In the previous section, we observed that Adsorption is quite effective at the class-instance acquisition task. However, a detailed comparison among different graph SSL algorithms is still missing. Talukdar and Pereira [2010] address this gap and present comparisons among three graph-based SSL algorithms: Gaussian Random Fields (referred to as LP_ZGL in these experiments), Adsorption, and Modified Adsorption (MAD). MRR is again used the evaluation metric in all experiments.

In the first set of experiments, Talukdar and Pereira [2010] use a subset of the Wikipedia-derived class-instance data made available by Pantel et al. (downloadable from http://ow.l y/13B57) for seeding and evaluation. This set consisted of 23 classes. Experiments comparing the three graph SSL methods using Freebase-1 graph (above) and this seed and evaluation is presented in Figure 5.7. From this figure, we observe that while LP-ZGL and Adsorption perform comparably, MAD significantly outperforms both these techniques. In order to evaluate how the algorithms scale up, Talukdar and Pereira [2010] experiment with a larger graph (Freebase-2) using 192 Wordnet classes, and observed similar performance pattern as above.

One natural question to consider is: in what settings is MAD likely to be particularly effective (e.g., as in Figure 5.7)? By summarizing experimental results from multiple settings, Talukdar and Pereira [2010] observe that MAD is likely to be most effective when applied to graphs with average high degree. This might be attributed to the additional regularization in MAD, which is missing in under-regularized methods such as GRF. Please see Talukdar and Pereira [2010] for additional experiments and further discussion.

Table 5.7: Inputs to PIDGIN [Wijaya et al., 2013] include KB$_1$ and KB$_2$, each consisting of two relation instances (e.g., (*Bill_Clinton, bornIn, Hope*)). Another input is a set of Subject-Verb-Object (SVO) triples (the interlingua) extracted from a natural language text corpus. Table is reproduced from Wijaya et al. [2013]

Knowledge Base 1 (KB$_1$):
(Rihanna, bornIn, St. Michael)
(Bill_Clinton, bornIn, Hope)

Knowledge Base 2 (KB$_2$):
(Reagan, personBornInCity, Tampico)
(Obama, personBornInCity, Honolulu)

Interlingua (Subject-Verb-Object):
(Bill Clinton, was born in, Hope)
(Barack Obama, was born in, Honolulu)
. . .

5.5   KNOWLEDGE BASE ALIGNMENT

Over the last few years, several large, publicly available Knowledge Bases (KBs) have been constructed. They include DBPedia [Auer et al., 2007], Freebase [Bollacker et al., 2008], NELL [Carlson et al., 2010], and Yago [Suchanek et al., 2007]. These KBs consist of both an ontology that defines a set of categories (e.g., *Athlete, Sports*) and relations (e.g., *playerPlaysSport(Athlete, Sport)*), and the data entries which instantiate these categories (e.g., *Tiger Woods* is an *Athlete*) and relations (e.g., *playerPlaysSport(Tiger Woods, Golf)*). The growth of the Semantic Web [Berners-Lee et al., 2001] has contributed significantly to the construction and availability of such KBs. However, these KBs are often independently developed, using different terminologies, coverage, and ontological structure of categories and relations. Therefore, the need for automatic alignment of categories and relations across these and many other heterogeneous KBs is now greater than ever, and this remains one of the core unresolved challenges of the Linked data movement [Bizer et al., 2009].

Research within the ontology matching community has addressed different aspects of the alignment problem, with recently proposed PARIS [Suchanek et al., 2011] being the current state-of-the-art in this large body of work (see Shvaiko and Euzenat [2012], for a recent survey). PARIS is a probabilistic ontology matcher which uses the overlap of instances between two relations (or categories) from a pair of KBs as one of the primary cues to determine whether an equivalence or subsumption relationship exists between those two relations (or categories). PARIS, and the instance overlap principle which it shares with most previous ontology alignment systems, has been found to be very effective in discovering alignments when applied to KBs such as DBPedia [Auer et al., 2007], Yago [Suchanek et al., 2007], and IMDB.[1]

---

[1]http://www.imdb.com/

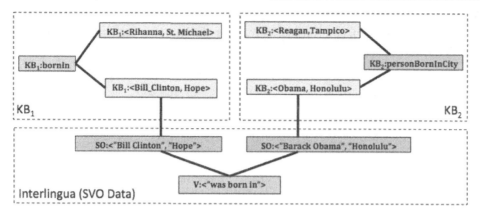

Figure 5.8: Graph constructed by PIDGIN from the input in Table 5.7. PIDGIN performs inference over this graph using Modified Adsorption (Section 3.2.5) to determine that KB_1:bornIn is equivalent to KB_2:personBornInCity. Note since there is no overlap between relation instances from these two KBs, algorithms based on instance overlap will be unable to align these two ontologies. PIDGIN overcomes this limitation through use of the SVO-based interlingua and inference over the graph. Figure is reproduced from Wijaya et al. [2013].

Despite this recent progress, the current state-of-the-art remains insufficient to align ontologies across many practical KB's and databases, especially when they share few or no data entries in common. To overcome this shortcoming, Wijaya et al. [2013] introduce the PIDGIN system which is capable of matching the categories and relations across multiple KB's even in the extreme case where they share no data entries in common. The key idea is to introduce side information in the form of a very large text corpus (in this case, 500 million dependency-parsed web pages). PIDGIN, effectively grounds each KB relation instance (e.g., *playerPlaysSport(Rodriguez, baseball)*) by its mentions in this text, then represents the relation in terms of the verbs that connect its arguments (e.g., the relation *playerPlaysSport(x,y)* might frequently be expressed in text by verbs such as "*x plays y*" or "*x mastered y*"). The distribution of verbs associated with instances of any given relation forms a KB-independent representation of that relation's semantics, which can then be aligned with relations from other KBs. In essence, the verb distributions associated with relations provide an *interlingua* that forms the basis for aligning ontologies across arbitrary KBs, even when their actual data entries fail to overlap. PIDGIN integrates this text information with information about overlapping relation instances across the two KB's using Modified Adsorption (MAD), to determine their final ontology alignment. PIDGIN consists of two stages:

- **Graph Construction**: Given two KBs and a Subject-Verb-Object (SVO) based interlingua, PIDGIN first represents this data as a graph. An example graph constructed from sample input is shown in Figure 5.8.

Table 5.8: Statistics of KBs used in experiments. In the experiments, NELL is used as a common target to align other KBs to, and consider only those relations in other KBs that have alignments to NELL relations (as decided by human annotators). Table is reproduced from Wijaya et al. [2013]

KB	Relations	Relation Instances
Freebase	79	7,450,452
NELL	499	3,235,218
Yago2	23	1,770,163
KBP	17	1,727

- **Alignment as Classification over Graph**: Once the graph is constructed, PIDGIN poses ontology alignment as node classification over this graph. For the graph in Figure 5.8, PIDGIN first associates two labels with the node KB_1:*bornIn*, one for equivalence and the other for subsumption, both specific to this node. For each relation r_1 in KB_1, PIDGIN generates two labels and injects them as seed labels into nodes of the graph G as follows.

  - l_{r_1}: This label is node-specific, and is injected *only* on the node named KB_1:r_1 which corresponds to relation r_1 in V, i.e., this is *self injection*. This label will be used to establish equivalence with other relations in KB_2.

  - $l_{\bar{r}_1}$: This label is injected as seed nodes corresponding to children of relation r_1 as determined by ontology of KB_1. However, if no such child exists, then this label is effectively discarded. This label will be used to identify subsumption relations in KB_2 which are subsumed by, i.e., less general than, r_1.

Starting with this initial seed information and graph structure, PIDGIN uses a graph-based semi-supervised learning (SSL) algorithm to classify the rest of the nodes in the graph, including the node corresponding to the relation in KB_2. Based on the assignment of scores of these labels on the KB_2 relation node, PIDGIN will determine the alignments between ontologies from these two KBs. PIDGIN starts out by attempting to align relations from the two KBs, and produces category alignment as an important by product.

Please note that PIDGIN is *self-supervised*, and hence doesn't require any additional human supervision. Similarly ideas were also explored for the problem of schema matching in [Talukdar et al., 2010]. Further for ease of explanation and readability, all examples and descriptions here involve two categories. However, PIDGIN is capable of handling multiple ontologies simultaneously.

DATASETS AND RESULTS

Table 5.8 shows the statistics of the KBs used in Wijaya et al. [2013]. They include several large scale open-domain publicly available real-world KBs, namely NELL [Carlson et al., 2010] (a

Table 5.9: Precision, Recall, and F1 scores @k=1 of Relation equivalence alignments comparing overlap based approach such as JACCARD and PARIS with PIDGIN. For each KB pair, best performance is marked in bold. Table is reproduced from Wijaya et al. [2013]

KB Pair	System	Prec	Recall	F1
	JACCARD (inst)	0.61	0.51	0.56
Freebase & NELL	PARIS	0.47	0.09	0.15
	PIDGIN	**0.65**	**0.61**	**0.63**
	JACCARD (inst)	0.56	0.43	0.49
Yago2 & NELL	PARIS	**0.67**	0.09	0.15
	PIDGIN	0.52	**0.52**	**0.52**
	JACCARD (inst)	0.0	0.0	0.0
KBP & NELL	PARIS	0.0	0.0	0.0
	PIDGIN	**0.07**	**0.06**	**0.06**

Table 5.10: Precision, Recall, and F1 scores @k=1 of relation subsumption alignments comparing PARIS with PIDGIN. For each KB pair, best performance is marked in bold. Table is reproduced from Wijaya et al. [2013]

KB Pair	System	Prec	Recall	F1
	PARIS	0.36	0.08	0.13
Freebase & NELL	PIDGIN	**0.8**	**0.77**	**0.79**
	PARIS	0.33	0.06	0.09
Yago2 & NELL	PIDGIN	**0.65**	**0.61**	**0.63**
	PARIS	1.0	0.13	0.24
KBP & NELL	PIDGIN	**0.8**	**0.8**	**0.8**

large-scale KB extracted automatically from web text), Yago2 [Hoffart et al., 2011] (a large-scale KB extracted automatically from semi-structured text of Wikipedia infoboxes), Freebase [Bollacker et al., 2008] (a large-scale KB created collaboratively and manually by humans), and KB Population (KBP) dataset (a smaller scale, manually constructed dataset used in the 2012 Text Analysis Conference for entity-linking, slot-filling and KB population tasks[2]).

Experimental results comparing PIDGIN to two other baselines: one that uses the Jaccard Similarity measure over instance overlap, and another state-of-the-art ontology alignment system PARIS [Suchanek et al., 2011], are shown in Table 5.9 (relation equivalence) and Table 5.10 (relation subsumption). From these results, we observe that PIDGIN gives the best performance in all settings. This demonstrates benefits of PIDGIN's use of natural language web text corpus as interlingua and a graph-based self-supervised learning to infer alignments.

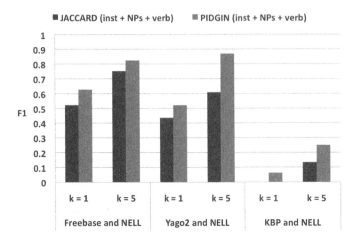

Figure 5.9: F1 scores @*k* = 1 and 5 of relation equivalence alignments comparing overlap-based approach (black) and PIDGIN (grey) both using the same set of resources. This demonstrates the benefit of transitivity of inference which is exploited by PIDGIN. Figure is reproduced from Wijaya et al. [2013].

Effect of Transitivity of Inference

Here, Wijaya et al. [2013] evaluate whether the transitivity of inference and joint inference in label propagation (PIDGIN) improve alignment performance when compared to the overlap-based approach (JACCARD). We observe in Figure 5.9 that using PIDGIN improves performance (F1-scores) at various values of *k* compared to the overlap-based approach even when the same set of resources are being used in both. The joint inference framework of PIDGIN that jointly optimizes similarities coming from instances, NP pairs and verbs, may be what makes it more tolerant to noise in the interlingua than a simple overlap-based similarity.

Learning Verbs for Relations

As a by-product of label propagation on the graph, each verb and NP-pair node in the graph will be assigned scores for each relation label. Exploiting these scores, we can estimate the probability that a verb *v* represents a relation *r* (see Wijaya et al. [2013] for details). Using this scoring, for each relation we can return a ranked list of verbs that represent the relation. Some of the verbs returned are shown in Table 5.11. As we can see in Table 5.11, the system is able to distinguish verbs representing the relation */medicine/medical_treatment/side_effects*: "*exacerbate*," "*can cause*" from the verbs representing the antonym relation *drugpossiblytreatsPossiblyTreatsPhysiologicalCondition*: "*relieve*," "*can help alleviate*" even when the two relations have the same domain (*drug*) and range (*physiological condition*). The system is also able to recognize the directionality

[2]http://www.nist.gov/tac/2012/KBP/index.html

Table 5.11: Examples of relation-verb pairs automatically learned by PIDGIN. Although verb stems were used in the experiments, for better readability, the original non-stemmed forms of the same verbs are shown. Table is reproduced from Wijaya et al. [2013]

Knowledge Base	Relation	Learned Verbs
Freebase	/sports/sports_team/ arena_stadium	played at, played in, defeated at, will host at, beaten at
	/medicine/medical_ treatment/side_effects	may promote, can cause, may produce, is worsen, exacerbate
NELL	drugPossiblyTreats PhysiologicalCondition	treat, relieve, reduce, help with, can help alleviate
	politicianHoldsOffice	serves as, run for, became, was elected
Yago2	actedIn	played in, starred in, starred, played, portrayed in
	isMarriedTo	married, met, date, wed, divorce

of the relation. For example, for the relation _acquired, which represents the inverse of the relation *acquired* (as in company X acquiring company Y); the system is able to return the correct verbs: _bought and _purchase, which are the inverse forms of the verbs *bought* and *purchase* (as in *is bought by* and *is purchased by*). Of practical importance is the fact that PIDGIN can learn verbs representing relations in Freebase and Yago2 whose facts are created manually or extracted via carefully constructed regular-expression matching. We can now use these verbs to automate an extraction process for the ontologies used by Freebase and Yago2. Please see Section 5 of Wijaya et al. [2013] for additional experiments evaluating PIDGIN's performance.

5.6 CONCLUSION

In this chapter, we reviewed at a variety of applications where graph-based SSL techniques have been successfully applied. In Section 5.1, we looked at the problem of text classification. From the experiments, we find that all the SSL approaches outperform supervised baseline. Moreover, the graph-based SSL algorithms perform better than non graph-based algorithms. In particular, the MAD algorithm achieved the best performance. Then in Section 5.2, we consider the problem of phone classification, an important step towards speech recognition. Here also we find that graph-based SSL techniques, especially Measure Propagation (MP), are able to make effective use of available unlabeled data and significantly outperform state-of-the-art supervised methods. In Section 5.3, we look at the unique setting where graph-based SSL is used to generate features for a downstream structured predictor, instead of using the graph-based SSL algorithm directly for prediction, as in most other applications. Here, we find that performance improves significantly when the structured predictor is infused with resources generated by graph-based

SSL. In Section 5.4, we observe how graph-based SSL techniques may be used to integrate class-instance tuples extracted by different methods and from different resources. Finally, in Section 5.5 we review how graph-based SSL may be applied to the problem of aligning ontologies from different KBs.

This is only a partial sample of graph-based SSL applications, and there are many more. For example, in Das and Petrov [2011], graph-based SSL is used for multilingual PoS tagging. In Velikovich et al. [2010], graph SSL is used to extent lexicons for sentiment classification. In Alexandrescu and Kirchhoff [2009], graph-based SSL is used for applications in Statistical Machine Translation (SMT).

CHAPTER 6

Future Work

In this chapter, we outline some directions for future work in each of the main components of graph-based SSL, namely, graph construction, inference and scalability.

6.1 GRAPH CONSTRUCTION

In spite of its importance, graph construction for graph-based SSL has not received adequate attention.

- **Regular vs. Irregular Graphs**: Jebara et al. [2009] emphasize the importance of regular graphs for graph-based SSL compared to irregular graphs which are often the result of graph-construction methods such as k-NN. However, the relationship between node degrees and performance on a classification task is not entirely obvious. In *under-regularized* methods, such as Zhu et al. [2003], it is conceivable that the presence of a high degree node may have an adverse effect as the algorithm may end up assigning most of the nodes the same label as a high degree node, thereby producing a degenerate solution. However, recent graph-based SSL algorithms [Baluja et al., 2008, Talukdar and Crammer, 2009] have attempted to address this problem by decreasing importance of high-degree nodes through additional regularization. In the light of these developments, it is worth exploring whether the computationally expensive algorithms for constructing regular graphs can be replaced by k-NN based graph construction while high-degree nodes are regularized during inference.

- **Construction based on Global Property**: All the graph construction methods presented in Chapter 2 emphasize local properties of the graph. For example, each node should have a certain number of neighbors. However, there has not been much work in the direction of enforcing other local properties or even global properties, e.g., constraints on the diameter of the graph. Optimizing over a global property will most likely give rise to a combinatorial explosion and so certain efficiently enforceable approximations of the global property may have to be considered.

- **Constructing Directed Graphs**: Almost all of the work in graph construction has been in inducing undirected graphs. This is primarily because graph-based transductive methods are targeted towards undirected graphs. However, recently transductive methods for directed graphs have been proposed [Subramanya and Bilmes, 2010, Zhou et al., 2005] and so there is scope for induction of graphs with non-symmetric edge weight matrices.

- **Alternative Graph Structures**: All graph structures considered so far have instance-instance edges. There has been very little work in learning other types of graphs, e.g., hybrid graphs with both features and instances present in them. Such types of graphs are used in Wang, Song, and Zhang [2009], although the graph structure in this case is fixed a-priori and is not learned.

- **Including Prior Information**: All the graph construction methods surveyed so far [Daitch et al., 2009, Jebara et al., 2009, Wang and Zhang, 2008] use the locally linear reconstruction (LLR) principle [Roweis and Saul, 2000] in one way or the other. For example, Daitch et al. [2009] and Wang and Zhang [2008] use variants of LLR during construction of the graph itself, while LLR is used for edge re-weighting in Jebara et al. [2009]. This suggests that there is room for exploration in this space in terms of other construction heuristics and methods. For example, none of these methods take prior knowledge into account. Sometimes we may know *a priori* the existence of certain edges and in the resulting graph we would like to include those edges. Such partial information may regularize the construction method and result in more appropriate graphs for the learning task at hand.

- **Task-dependent Graph Construction**: This line of more exploration deserves more attention and in our opinion holds the greatest promise of impact. As we saw in Section 2.3, incorporation of task-specific information can lead to significant improvement in performance, but there is lot of room for exploration and improvement. Using multiple kernels instead of one during graph construction is one obvious area of future work. In general, connection between graph construction and metric learning needs to be more thoroughly explored.

6.2 LEARNING & INFERENCE

- **Degenerate Solutions**: Graph-based SSL methods sometimes lead to degenerate solutions, i.e., where all nodes in the graph are assigned the same label. Zhu et al. [2003] proposed using the *class mass normalization* (CMN) heuristic to solve this problem. Here, the inference decisions for the unlabeled points are post-hoc modified so that the output class proportions are similar to the input class proportions. The input class proportions are either estimated from the labeled data or using domain knowledge. The downside of CMN is that decisions made during inference are modified as a post-processing step and not integrated with the inference process itself. Note that in the case of probabilistic approaches such as MP and IR, it is possible to integrated CMN with the inference [Subramanya and Bilmes, 2010]. However, it is more challenging in the case of non-probabilistic approaches such as MAD or Adsorption.

  As we saw in Section 2.3.2, the eigenvectors of a graph Laplacian corresponding to the smaller eigenvalues are smoother and in particular cases constant across each connected component, an interesting research direction is to explore the relationship between the spectrum of the Laplacian and the possibility for degenerate solutions. For a given graph-based SSL algorithm, is it possible to know in advance the types of graphs on which it is likely to lead to degenerate solutions so that more principled corrective steps can be taken. Transformation of the Laplacian may be one possibility.

- **Directed Graphs**: Barring a few exceptions [Subramanya and Bilmes, 2010, Zhou et al., 2005], most of the literature on graph-based SSL is biased towards undirected graphs. This is not sufficient as many natural graphs (e.g., the hypertext graph, follower-followee graph in social networks, etc.) are directed in nature. More research is needed towards developing algorithms which can naturally handle such directed graphs.

- **Going beyond Node Similarity**: Almost all graph-based SSL techniques can only handle graphs with a single edge type, i.e., node similarity. This is extended to two edge types in GraphOrder [Wijaya et al., 2013], but that is not sufficient in many applications. Many natural graphs of practical significance consist of heterogeneous edge types (e.g., knowledge graphs [Bollacker et al., 2008, Carlson et al., 2010, Suchanek et al., 2007] consists of hundreds of different types of edges). More work is needed in the direction of graph-based SSL algorithms that can handle a wide diversity of edge types.

- **Structured Prediction Problems**: Apart from the work of Subramanya et al. [2010], Das and Petrov [2011], and He et al. [2013], there has been very little work in the direction of using graph-based methods for structured prediction problems. A number of problems of interest in speech, natural language, and vision have an inherent structure associated with the output space and annotating data for fully-supervised solutions of these problems is time-consuming and cumbersome and so they can benefit from the use of unlabeled data.

- **Joint Graph Construction and Inference**: As we have seen, each graph-based SSL algorithm makes certain assumptions about the underlying graph. As a result, it would be useful for graph construction schemes to take into account the types of graphs preferred by the inference algorithm. Further, it seems useful for graph construction to be based on minimizing the test error or some function of the test error.

6.3 SCALABILITY

- **Scalable Graph Construction**: Even though scalable graph construction has received some attention [Goyal et al., 2012], this is carried out in the task-independent setting. Scalable graph construction, especially in the task-dependent setting, is an important area for future work.

- **Scaling to Large Number of Labels in Non-Squared Loss Setting**: In Section 4.3, we observed how sketches may be used to scale graph-based SSL to settings with a large number of labels. However, this is only applicable (and effective) when the square loss is used in the graph SSL objective. Extending such ideas to settings with non-squared loss (e.g., [Subramanya and Bilmes, 2010]) is an interesting direction for future work.

APPENDIX A

Notations

In this section, we outline the notations used throughout the book.

- Let X be the $d \times n$ matrix of n instances in a d-dimensional space. Out of the n instances, n_l instances are labeled, while the remaining n_u instances are unlabeled, with $n = n_l + n_u$. We shall represent the ith instance (i.e., the ith column of X) by $\mathbf{x}_i \in \mathcal{R}^d$.

- Let $G = (V, E, W)$ be the graph constructed out of the instances in X, where V is the set of nodes with $|V| = n$, E is the set of edges, and W is the edge weight matrix. In other words, node $v_i \in V$ corresponds to instance x_i. Out of the $n = n_l + n_u$ nodes in G, n_l nodes are labeled, while the remaining n_u nodes are unlabeled.

- Unless specified otherwise, $W \in \mathcal{R}_+{}^{n \times n}$ is the symmetric matrix of edge weights. W_{ij} is the weight of edge (i, j), and also the similarity between instances x_i and x_j. If edge $(u, v) \notin E$, $W_{uv} = 0$. Further, there are no self-loops, i.e., $W_{uu} = 0$.

- The (unnormalized) Laplacian, L, of G is given by $L = D - W$, where D is an $n \times n$ diagonal degree matrix with $D_{uu} = \sum_v W_{uv}$.

- Let S be an $n \times n$ diagonal matrix with $S_{uu} = 1$ iff node $u \in V$ is labeled. That is, S identifies the labeled nodes in the graph.

- $\mathcal{Y}$ is the set of labels, with $|\mathcal{Y}| = m$ representing the total number of labels.

- Y is the $n \times m$ matrix storing training label information, if any.

- $\hat{Y}$ is an $n \times m$ matrix of soft label assignments, with $\hat{Y}_{vl}$ representing the score of label l on node v. A graph-based SSL computes $\hat{Y}$ from $\{G, SY\}$.

APPENDIX B

Solving Modified Adsorption (MAD) Objective

In this section, we look at details of the optimization carried out by the MAD algorithm (Section 3.2.5). We first reproduce MAD's objective from Equation 3.3:

$$\mathbf{C}^{(MAD)}(\hat{Y}) = \sum_{l} \left[\mu_1 \left(Y_l - \hat{Y}_l \right)^{\top} S \left(Y_l - \hat{Y}_l \right) + \mu_2 \hat{Y}_l^{T} L \hat{Y}_l + \mu_3 \left\| \hat{Y}_l - R_l \right\|^2 \right].$$

This objective can be solved using the same techniques as in the case of the quadratic cost criteria [Bengio et al., 2007]. The gradient of $\mathbf{C}^{(MAD)}(\hat{Y})$ is given by

$$\frac{1}{2} \frac{\delta C(\hat{Y})}{\delta \hat{Y}_l} = (\mu_1 S + \mu_2 L + \mu_3 I)\hat{Y}_l - (\mu_1 S Y_l + \mu_3 R_l). \tag{B.1}$$

The second derivative is given by

$$\frac{1}{2} \frac{\delta^2 C(\hat{Y})}{\delta \hat{Y}_l \delta \hat{Y}_l} = \mu_1 S + \mu_2 L + \mu_3 I$$

and as both S and L are symmetric and positive semidefinite matrices (PSD), we get that the Hessian (Equation (B.2)) is PSD as well, since positive semi-definiteness is preserved under non-negative scalar multiplication and addition. Hence, we get the optimal minima is obtained by setting the first derivative (i.e., Equation (B.1)) to 0, and we get,

$$\hat{Y}_l = [\mu_1 S + \mu_2 L + \mu_3 I]^{-1} (\mu_1 S Y_l + \mu_3 R_l).$$

Computing the labels ($\hat{Y}$) requires both a matrix inversion and matrix multiplication. This can be quite expensive when large matrices are involved. A more efficient way to obtain the new label scores is to solve a set of linear equations using Jacobi iteration [Saad, 2003].

Jacobi Method: Given a linear system in x

$$Mx = b,$$

the Jacobi iterative algorithm defines the approximate solution at the $(t + 1)^{th}$ iteration as

$$x_i^{(t+1)} = \frac{1}{M_{ii}} \left(b - \sum_{j \neq i} M_{ij} x_j^{(t)} \right), \tag{B.2}$$

where $x_j^{(t)}$ is the solution at iteration t. We apply the iterative algorithm to our problem by substituting $x = \hat{Y}_l$, $M = \mu_1 S + \mu_2 L + \mu_3 I$ and $b = \mu_1 S Y_l + \mu_3 R_l$ in (B.2) and so

$$\hat{Y}_{vl}^{(t+1)} = \frac{1}{M_{vv}} \left(\mu_1 (S Y_l)_v + \mu_3 R_{vl} - \sum_{u \neq v} M_{vu} \hat{Y}_{ul}^{(t)} \right).$$

(B.3)

Observe that

$$M_{vu} = \mu_1 S_{vu} + \mu_2 L_{vu} + \mu_3 I_{vu} \; \forall u \neq v.$$

As S and I are diagonal matrices, we have that $S_{vu} = 0$ and $I_{vu} = 0$ for all $u \neq v$ and so we have that

$$M_{vu(v \neq u)} = \mu_2 L_{vu} = \mu_2 \left(D_{vu} + \bar{D}_{vu} - W'_{vu} - W'_{uv} \right),$$

(B.4)

and as before the matrices D and $\bar{D}$ are diagonal and thus $D_{vu} + \bar{D}_{vu} = 0$. Finally, substituting the values of W'_{vu} and W'_{uv} we get

$$M_{vu(v \neq u)} = -\mu_2 \times (p_v^{cont} \times W_{vu} + p_u^{cont} \times W_{uv}).$$

(B.5)

We now compute the second quantity,

$$(S Y_l)_{vu} = S_{vv} Y_{vv} + \sum_{t \neq v} S_{vt} Y_{tu} = p_v^{inj} \times Y_{vv},$$

where the second term in the first line equals zero since S is diagonal. Finally, the third term,

$$\begin{aligned} M_{vv} &= \mu_1 S_{vv} + \mu_2 L_{vv} + \mu_3 I_{vv} \\ &= \mu_1 \times p_v^{inj} + \mu_2 (D_{vv} - W'_{vv}) + \mu_3 \\ &= \mu_1 \times p_v^{inj} + \mu_2 \times p_v^{cont} \times \sum_{u \neq v} W_{vu} + \mu_3. \end{aligned}$$

Plugging the above equations into (B.3) and using the fact that the diagonal elements of W are zero, we get

$$\hat{Y}_v^{(t+1)} = \frac{1}{M_{vv}} \left(\mu_1 p_v^{inj} Y_v + \mu_2 \sum_u \left(p_v^{cont} W_{vu} + p_u^{cont} W_{uv} \right) \hat{Y}_u^{(t)} + \mu_3 p_v^{abnd} R_v \right).$$

(B.6)

This is the resulting iterative update used by MAD, as shown in Algorithm 3.

APPENDIX C

Alternating Minimization

Given a distance function $d(a, b)$ where $a \in \mathcal{A}, b \in \mathcal{B}$ and $\mathcal{A}, \mathcal{B}$ are sets, consider the problem finding the a^*, b^* that minimizes $d(a, b)$. AM is a very effective tool for optimizing such functions. AM refers to the case where we alternately minimize $d(p, q)$ with respect to p while q is held fixed and then vice versa, that is,

$$a^{(n)} = \underset{a \in \mathcal{A}}{\operatorname{argmin}}\, d(a, b^{(n-1)}) \text{ and } b^{(n)} = \underset{b \in \mathcal{B}}{\operatorname{argmin}}\, d(a^{(n)}, b).$$

Figure C.1 illustrates the two steps of AM over two convex sets. It starts with an initial arbitrary $b_0 \in \mathcal{B}$ which is held fixed while we minimize w.r.t. $a \in \mathcal{A}$ which leads to a_1. The objective is then held fixed w.r.t. $\mathcal{A}$ at $a = a_1$ and minimized over $b \in \mathcal{B}$ and this leads to b_1. The above is then repeated with b_1 playing the role of b_0 and so on until (in the best of cases) convergence. The Expectation-Maximization (EM) [Dempster et al., 1977] algorithm is an example of AM. Moreover, the above objective over two variables can be extended to an objective over n variables. In such cases $n - 1$ variables are held fixed while the objective is optimized with respect to the one remaining variable and the procedure iterates in a similar round-robin fashion.

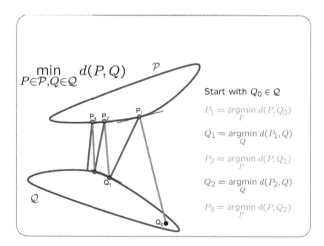

Figure C.1: Illustration of alternating minimization. Figure reproduced from [Subramanya and Bilmes, 2010].

The convergence properties of AM depend on $d(a, b)$ and the nature of domain of a and b. Even in situations where AM does converge, it may not converge to the true minimum of $d(a, b)$. In general, certain conditions need to hold for AM to converge to the correct solution. Some approaches, such as Cheney and Goldstien [1959] and Wu [1983], rely on the topological properties of the objective and the space over which it is optimized, while others such as [Csiszar and Tusnady, 1984] use geometrical arguments. Still others [Gunawardena, 2001] use a combination of the above.

APPENDIX D

Software

In this appendix, we provide an overview of Junto, a graph-based SSL toolkit available for download from `https://github.com/parthatalukdar/junto`. For additional software listings including non-graph SSL techniques, please refer to Appendix B of Zhu and Goldberg [2009].

D.1 JUNTO LABEL PROPAGATION TOOLKIT

The Junto toolkit provides implementations of three graph-based SSL algorithms: GRF (referred to as LP-ZGL in the toolkit) [Zhu et al., 2003], Adsorption [Baluja et al., 2008], and Modified Adsorption (MAD) [Talukdar and Crammer, 2009]. Both stand alone, as well as Hadoop-based distributed implementations of these algorithms, are available in the toolkit. Junto is written in a mix of Scala and Java. Please refer to the README in the toolkit for installation details.

The conjuration keys and their values may be specified in a configuration file or specified on the command line itself. If a config_key is present in both the config file and the command line, then the latter overwrites the former. After installation is complete, Junto may be invoked as follows:

```
$ junto config config_file [config_key=value ...]
```

Please look into sample config files such as `examples/simple/simple_config` in the toolkit for annotated description of various configurations supported by the algorithm. If requested, predictions from the algorithms are stored in the file specified using configuration option `output_file`. By default, the toolkit reports precision and MRR (Mean Reciprocal Rank) whenever evaluation data is supplied.

To use the toolkit in Hadoop mode, please see `examples/hadoop/README`, with sample configuration file provided in `examples/hadoop/simple_hadoop_config`.

Bibliography

A. Alexandrescu and K. Kirchhoff. Graph-based learning for phonetic classification. In *Proceedings of the Automatic Speech Recognition and Understanding Workshop (ASRU)*, 2007. DOI: 10.1109/ASRU.2007.4430138. 66

A. Alexandrescu and K. Kirchhoff. Graph-based learning for statistical machine translation. In *Proceedings of the Conference of Human Language Technologies-North American Association for Computational Linguistics (HLT-NAACL)*, 2009. 1, 84

Y. Altun, D. McAllester, and M. Belkin. Maximum margin semi-supervised learning for structured variables. In *Proceedings of the Advances in Neural Information Processing Systems (NIPS)*, 2005. 69

A. Andoni and P. Indyk. Near-optimal hashing algorithms for approximate nearest neighbor in high dimensions. *Communications of the ACM*, 2008. DOI: 10.1145/1327452.1327494. 48, 58

S. Arya and D. M. Mount. Approximate nearest neighbor queries in fixed dimensions. In *Proceedings of the ACM-SIAM Symposium on Discrete Algorithms (SODA)*, 1993. 48, 57, 64

S. Arya, D. M. Mount, N. S. Netanyahu, R. Silverman, and A. Wu. An optimal algorithm for approximate nearest neighbor searching. *Journal of the ACM*, 1998. DOI: 10.1145/293347.293348. 48, 58, 64

S. Auer, C. Bizer, G. Kobilarov, J. Lehmann, R. Cyganiak, and Z. Ives. Dbpedia: A nucleus for a web of open data. In *The semantic web*. 2007. DOI: 10.1007/978-3-540-76298-0_52. 78

S. Baluja, R. Seth, D. Sivakumar, Y. Jing, J. Yagnik, S. Kumar, D. Ravichandran, and M. Aly. Video suggestion and discovery for YouTube: Taking Random Walks Through the view graph. In *Proceeding of the 17th international conference on World Wide Web (WWW)*, 2008. DOI: 10.1145/1367497.1367618. 29, 30, 43, 44, 85, 95

M. Belkin, P. Niyogi, and V. Sindhwani. On manifold regularization. In *Proceedings of the Conference on Artificial Intelligence and Statistics (AISTATS)*, 2005. 13, 20, 40, 41, 42, 43, 45

Y. Bengio, O. Delalleau, and N. L. Roux. *Semi-Supervised Learning*, chapter Label Propogation and Quadratic Criterion. MIT Press, 2007. 34, 37, 43, 71, 91

J. Bentley. Multidimensional divide-and-conquer. *Communications of the ACM*, 23:214–229, 1980. DOI: 10.1145/358841.358850. 11

J. L. Bentley. Multidimensional binary search trees used for associative searching. *Communications of ACM*, 1975. DOI: 10.1145/361002.361007. 48, 57

T. Berners-Lee, J. Hendler, O. Lassila, et al. The semantic web. *Scientific american*, 2001. DOI: 10.1038/scientificamerican0501-34. 78

D. Bertsekas. *Nonlinear Programming*. Athena Scientific Publishing, 1999. 39

A. Beygelzimer, S. Kakade, and J. Langford. Cover trees for nearest neighbor. In *Proceedings of the International Conference on Machine Learning (ICML)*, 2006. DOI: 10.1145/1143844.1143857. 11

M. Bilenko, S. Basu, and R. Mooney. Integrating constraints and metric learning in semi-supervised clustering. In *Proceedings of the International Conference on Machine Learning (ICML)*, 2004. DOI: 10.1145/1015330.1015360. 19, 20

J. Bilmes and A. Subramanya. *Scaling up machine learning: Parallel and distributed approaches*, chapter Parallel Graph-based Semi-Supervised Learning. Cambridge University Press, 2011. 50, 51, 52, 53

E. Bingham and H. Mannila. Random projection in dimensionality reduction: Applications to image and text data. In *Proceedings of the International Conference on Knowledge Discovery and Data Mining (KDD)*, 2001. DOI: 10.1145/502512.502546. 20

C. Bishop, editor. *Neural Networks for Pattern Recognition*. Oxford University Press, 1995. 65

C. Bizer, T. Heath, and T. Berners-Lee. Linked data-the story so far. *International journal on semantic web and information systems*, 5(3):1–22, 2009. DOI: 10.4018/jswis.2009081901. 78

J. Blitzer, K. Q. Weinberger, and L. K. Saul. Distance metric learning for large margin nearest neighbor classification. In *Proceedings of the Advances in Neural Information Processing Systems (NIPS)*, 2005. 21

J. Blitzer, R. McDonald, and F. Pereira. Domain adaptation with structural correspondence learning. In *Proceedings of the Conference on Empirical Methods in Natural Language Processing (EMNLP)*, 2006. 68

A. Blum and S. Chawla. Learning from labeled and unlabeled data using graph mincuts. In *Proceedings of the International Conference on Machine Learning (ICML)*, 2001. 27, 28, 38, 39, 44, 47

A. Blum and T. Mitchell. Combining labeled and unlabeled data with co-training. In *Proceedings of the Workshop on Computational Learning Theory (COLT)*, 1998. DOI: 10.1145/279943.279962. 5

A. Blum, J. D. Lafferty, M. R. Rwebangira, and R. Reddy. Semi-supervised learning using randomized mincuts. In *International Conference on Machine Learning (ICML)*, 2004. DOI: 10.1145/1015330.1015429. 28

K. Bollacker, C. Evans, P. Paritosh, T. Sturge, and J. Taylor. Freebase: a collaboratively created graph database for structuring human knowledge. In *Proceedings of the International Conference on the Management of Data (SIGMOD)*, 2008. DOI: 10.1145/1376616.1376746. 78, 81, 87

S. Boyd and L. Vandenberghe. *Convex optimization*. Cambridge university press, 2004. DOI: 10.1017/CBO9780511804441. 24

C. Burges. A tutorial on support vector machines for pattern recognition. *Data mining and knowledge discovery*, 2(2):121–167, 1998. DOI: 10.1023/A:1009715923555. 25

M. Cafarella, A. Halevy, D. Wang, E. Wu, and Y. Zhang. Webtables: Exploring the power of tables on the web. In *Proceedings of the International Conference on Very Large Databases (VLDB)*, 2008. DOI: 10.14778/1453856.1453916. 73, 74, 76

A. Carlson, J. Betteridge, B. Kisiel, B. Settles, E. R. Hruschka Jr, and T. M. Mitchell. Toward an architecture for never-ending language learning. In *Proceedings of the Conference on Artificial Intelligence (AAAI)*, 2010. 55, 78, 80, 87

O. Chapelle, J. Weston, and B. Scholkopf. Cluster kernels for semi-supervised learning. In *Proceedings of the Advances in Neural Information Processing Systems (NIPS)*, 2002. 23

O. Chapelle, B. Scholkopf, and A. Zien. *Semi-Supervised Learning*. MIT Press, 2007. 5, 42, 43

W. Cheney and A. Goldstien. Proximity maps for convex sets. *American Mathematical Society*, 1959. DOI: 10.1090/S0002-9939-1959-0105008-8. 94

F. Chung. *Spectral graph theory*. American Mathematical Society, 1997. 24

A. Corduneanu. *The Information Regularization Framework for Semi-Supervised Learning*. PhD thesis, Massachusetts Institute of Technology, 2006. 36, 38, 44

A. Corduneanu and T. Jaakkola. On information regularization. In *Proceedings of the Conference on Uncertainty in Artificial Intelligence (UAI)*, 2003. 36, 37, 43

G. Cormode and S. Muthukrishnan. An improved data stream summary: the count-min sketch and its applications. *Journal of Algorithms*, 55(1):58–75, 2005. DOI: 10.1016/j.jalgor.2003.12.001. 55, 56, 59

K. Crammer and Y. Singer. On the algorithmic implementation of multiclass kernel-based vector machines. *Journal of Machine Learning Research (JMLR)*, 2, Mar. 2002. 63

N. Cristianini, J. Kandola, and A. Elissee. On kernel target alignment. In *Proceedings of the Advances in Neural Information Processing Systems (NIPS)*, 2001. DOI: 10.1007/3-540-33486-6_8. 24

I. Csiszar and G. Tusnady. Information geometry and alternating minimization procedures. *Statistics and Decisions*, 1984. 94

S. Daitch, J. Kelner, and D. Spielman. Fitting a graph to vector data. In *Proceedings of the International Conference on Machine Learning (ICML)*, 2009. DOI: 10.1145/1553374.1553400. 9, 14, 15, 16, 26, 86

D. Das and S. Petrov. Unsupervised part-of-speech tagging with bilingual graph-based projections. In *Proceedings of the Annual Meeting of the Association for Computational Linguistics (ACL)*, 2011. 84, 87

D. Das and N. A. Smith. Graph-based lexicon expansion with sparsity-inducing penalties. In *Proceedings of the Conference of the North American Chapter of the Association for Computational Linguistics: Human Language Technologies (NAACL-HLT)*, 2012. 1

J. Davis, B. Kulis, P. Jain, S. Sra, and I. Dhillon. Information-theoretic metric learning. In *International Conference on Machine Learning (ICML)*, 2007. DOI: 10.1145/1273496.1273523. 17, 18, 20

J. Dean and S. Ghemawat. Mapreduce: simplified data processing on large clusters. *Communications of the ACM*, 51(1):107–113, 2008. DOI: 10.1145/1327452.1327492. 54

O. Delalleau, Y. Bengio, and N. L. Roux. Efficient non-parametric function induction in semi-supervised learning. In *Proceedings of the Conference on Artificial Intelligence and Statistics (AISTATS)*, 2005. 48

A. Dempster, N. Laird, D. Rubin, et al. Maximum likelihood from incomplete data via the EM algorithm. *Journal of the Royal Statistical Society. Series B (Methodological)*, 39(1):1–38, 1977. 4, 93

J. Deng, W. Dong, R. Socher, L.-J. Li, K. Li, and L. Fei-Fei. Imagenet: A large-scale hierarchical image database. In *Proceedings of the IEEE Conference on Computer Vision and Pattern Recognition (CVPR)*, 2009. DOI: 10.1109/CVPR.2009.5206848. 55

P. Dhillon, P. P. Talukdar, and K. Crammer. Inference-driven metric learning (idml) for graph construction. Technical Report MS-CIS-10-18, University of Pennsylvania, 2010. 10, 16, 18, 19, 21

O. Duda, P. Hart, and D. Stork. *Pattern Classification*. Wiley-Interscience, second edition, 2001. 3

S. Dumais, J. Platt, D. Heckerman, and M. Sahami. Inductive learning algorithms and representations for text categorization. In *Proceedings of the International Conference on Information and Knowledge Management (CIKM)*, 1998. DOI: 10.1145/288627.288651. 61

A. Errity and J. McKenna. An investigation of manifold learning for speech analysis. In *Proceedings of INTERSPEECH*, 2006. 63

J. Friedman, J. Bentley, and R. Finkel. An algorithm for finding best matches in logarithmic expected time. *ACM Transaction on Mathematical Software*, 3, 1977. DOI: 10.1145/355744.355745. 48, 58, 64

J. Garcke and M. Griebel. Semi-supervised learning with sparse grids. In *Proceedings of the ICML Workshop on Learning with Partially Classified Training Data*, 2005. 48

J. Godfrey, E. Holliman, and J. McDaniel. Switchboard: Telephone speech corpus for research and development. In *Proceedings of the IEEE International Conference on Acoustics, Speech, and Signal Processing*, volume 1, pages 517–520, San Francisco, California, March 1992. DOI: 10.1109/ICASSP.1992.225858. 50, 51, 64

A. Goyal, H. Daumé III, and R. Guerra. Fast large-scale approximate graph construction for nlp. In *Proceedings of the Joint Conference on Empirical Methods in Natural Language Processing and Computational Natural Language Learning (EMNLP-CoNLL)*, 2012. 49, 87

S. Greenberg. The Switchboard transcription project. Technical report, The Johns Hopkins University (CLSP) Summer Research Workshop, 1995. 64

A. Gunawardena. *The Information Geometri of EM variants for Speech and Image Process ing*. PhD thesis, Johns Hopkins University, 2001. 94

A. K. Halberstadt and J. R. Glass. Heterogeneous acoustic measurements for phonetic classification. In *Proc. Eurospeech '97*, Rhodes, Greece, 1997. 63

L. He, J. Gillenwater, and B. Taskar. Graph-Based Posterior Regularization for Semi-Supervised Structured Prediction. In *Proc. Conference on Computational Natural Language Learning (CoNLL)*, 2013. 87

M. Hearst. Automatic acquisition of hyponyms from large text corpora. In *Proceedings of the 14th International Conference on Computational Linguistics (COLING-92)*, Nantes, France, 1992. DOI: 10.3115/992133.992154. 72

J. Hoffart, F. Suchanek, K. Berberich, E. Lewis-Kelham, G. De Melo, and G. Weikum. Yago2: exploring and querying world knowledge in time, space, context, and many languages. In *WWW*, 2011. 81

D. W. Hosmer. A comparison of iterative maximum likelihood estimates of the parameters of a mixture of two normal distributions under three different types of sample. *Biometrics*, 1973. DOI: 10.2307/2529141. 5

B. Huang and T. Jebara. Loopy belief propagation for bipartite maximum weight b-matching. *Artificial Intelligence and Statistics (AISTATS)*, 2007. 14

T. Jebara, J. Wang, and S. Chang. Graph construction and b-matching for semi-supervised learning. In *Proceedings of the 26th Annual International Conference on Machine Learning*. ACM New York, NY, USA, 2009. DOI: 10.1145/1553374.1553432. 9, 12, 13, 14, 21, 26, 85, 86

R. Jin, S. Wang, and Y. Zhou. Regularized Distance Metric Learning: Theory and Algorithm. In *NIPS*, 2009. 20

T. Joachims. Transductive inference for text classification using support vector machines. 1999. 43, 61, 63

T. Joachims. Transductive learning via spectral graph partitioning. In *Proc. of the International Conference on Machine Learning (ICML)*, 2003. 28, 39, 43, 44, 47, 61, 63

R. Johnson and T. Zhang. Graph-based semi-supervised learning and spectral kernel design. *IEEE Transactions on Information Theory*, 54(1):275–288, 2008. DOI: 10.1109/TIT.2007.911294. 25

J. Judge, A. Cahill, and J. van Genabith. Questionbank: Creating a corpus of parse-annotated questions. In *Proceedings of the 21st International Conference on Computational Linguist ics and 44th Annual Meeting of the Association for Computational Linguistics*, 2006. 68

M. Karlen, J. Weston, A. Erkan, and R. Collobert. Large scale manifold transduction. In *International Conference on Machine Learning, ICML*, 2008. DOI: 10.1145/1390156.1390213. 5, 55

K. Kirchhoff and A. Alexandrescu. Phonetic classification using controlled random walks. In *INTERSPEECH*, 2011. 65

S. B. Kotsiantis. Supervised machine learning: A review of classification techniques. *Informatica*, 31:249–268, 2007. 3

J. D. Lafferty, A. McCallum, and F. C. N. Pereira. Conditional random fields: Probabilistic models for segmenting and labeling sequence data. In *Proceedings of the Eighteenth International Conference on Machine Learning*, 2001. 67

G. Lanckriet, N. Cristianini, P. Bartlett, L. El Ghaoui, and M. Jordan. Learning the kernel matrix with semidefinite programming. *The Journal of Machine Learning Research*, 5:27–72, 2004. 24, 25

K. F. Lee and H. Hon. Speaker independant phone recognition using hidden markov models. *IEEE Transactions on Acoustics, Speech and Signal Processing*, 37(11), 1989. DOI: 10.1109/29.46546. 64

K. Lerman, S. Blair-Goldensohn, and R. McDonald. Sentiment summarization: Evaluating and learning user preferences. In *Proceedings of the 12th Conference of the European Chapter of the Association for Computational Linguistics*, EACL '09. Association for Computational Linguistics, 2009. DOI: 10.3115/1609067.1609124. 1

D. Lewis et al. Reuters-21578. `http://www.daviddlewis.com/resources/testcollecti ons/reuters21578`, 1987. 61

W. Liu, J. He, and S.-F. Chang. Large graph construction for scalable semi-supervised learning. In *Proceedings of the 27th International Conference on Machine Learning (ICML-10)*, pages 679–686, 2010. 48

M. Maier, U. von Luxburg, and M. Hein. Influence of graph construction on graph-based clustering measures. *The Neural Information Processing Systems*, 22:1025–1032, 2009. 9

A. McCallum and W. Li. Early results for named entity recognition with conditional random fields, feature induction and web-enhanced lexicons. In *Proceedings of the Human Language Technologies–North American Association for Computational Linguistics (HLT-NAACL)*, 2003. DOI: 10.3115/1119176.1119206. 67

G. J. McLachlan and S. Ganesalingam. Updating a Discriminant Function on the basis of Unclassified data. *Communication in Statistics: Simulation and Computation*, 1982. DOI: 10.1080/03610918208812293. 5

T. M. Mitchell. *Machine Learning*. McGraw-Hill, Inc., New York, NY, USA, 1997. 3

B. Nadler, N. Srebro, and X. Zhou. Statistical analysis of semi-supervised learning: The limit of infinite unlabelled data. In *Advances in Neural Information Processing Systems (NIPS)*, 2010. 47, 66

G. Nigam. *Using unlabeled data to improve text classification*. PhD thesis, CMU, 2001. 5

K. Nigam, A. McCallum, S. Thrun, and T. Mitchell. Learning to classify text from labeled and unlabeled documents. In *Proceedings of the fifteenth national/tenth conference on Artificial intelligence (AAAI 1998)*. 61

J. Nocedal and S. Wright. *Numerical optimization*. Springer, 1999. DOI: 10.1007/b98874. 15

M. Orbach and K. Crammer. Graph-based transduction with confidence. In *Machine Learning and Knowledge Discovery in Databases (ECML)*, 2012. DOI: 10.1007/978-3-642-33486-3_21. 27, 34, 35, 36, 44, 61

P. Pantel, E. Crestan, A. Borkovsky, A. Popescu, and V. Vyas. Web-scale distributional similarity and entity set expansion. *Proceedings of EMNLP-09.* 77

M. F. Porter. Readings in information retrieval. chapter An Algorithm for Suffix Stripping. Morgan Kaufmann Publishers Inc., 1997. 62

S. Roweis and L. Saul. Nonlinear dimensionality reduction by locally linear embedding. *Science*, 290(5500):2323–2326, 2000. DOI: 10.1126/science.290.5500.2323. 13, 15, 16, 26, 86

Y. Saad. *Iterative Methods for Sparse Linear Systems.* Society for Industrial Mathematics, 2003. DOI: 10.1137/1.9780898718003. 33, 91

G. Salton and C. Buckley. Term weighting approaches in automatic text retrieval. Technical report, Cornell University, Ithaca, NY, USA, 1987. DOI: 10.1016/0306-4573(88)90021-0. 62

B. Scholkopf and A. J. Smola. *Learning with Kernels: Support Vector Machines, Regularization, Optimization, and Beyond.* MIT Press, 2001. 3

H. J. Scudder. Probability of Error of some Adaptive Pattern-Recognition Machines. *IEEE Transactions on Information Theory*, 11, 1965. DOI: 10.1109/TIT.1965.1053799. 4

J. Shi and J. Malik. Normalized cuts and image segmentation. In *Proceedings of the 1997 Conference on Computer Vision and Pattern Recognition (CVPR 1997).* DOI: 10.1109/34.868688. 28

J. Shi and J. Malik. Normalized cuts and image segmentation. *IEEE Transactions on Pattern Analysis and Machine Intelligence*, 2000. DOI: 10.1109/CVPR.1997.609407. 52

Q. Shi, J. Petterson, G. Dror, J. Langford, A. Smola, and S. Vishwanathan. Hash kernels for structured data. *The Journal of Machine Learning Research*, 10:2615–2637, 2009. DOI: 10.1145/1577069.1755873. 55

P. Shvaiko and J. Euzenat. Ontology matching: state of the art and future challenges. *TKDE*, 2012. DOI: 10.1109/TKDE.2011.253. 78

A. Smola and R. Kondor. Kernels and regularization on graphs. In *16th Annual Conference on Learning Theory and 7th Kernel Workshop, COLT/Kernel 2003, Washington, DC, USA*, 2003. DOI: 10.1007/978-3-540-45167-9_12. 23, 25

A. Subramanya and J. Bilmes. Soft-supervised learning for text classification. In *Empirical Methods in Natural Language Processing (EMNLP)*, 2008. DOI: 10.3115/1613715.1613857. 1, 61, 62, 63

A. Subramanya and J. Bilmes. Semi-supervised learning with measure propagation. *Journal of Machine Learning Research*, 2010. 1, 6, 9, 27, 38, 39, 40, 43, 45, 64, 65, 66, 85, 86, 87, 88, 93

A. Subramanya, S. Petrov, and F. Pereira. Efficient graph-based semi-supervised learning of structured tagging models. In *Proceedings of the Conference on Empirical Methods in Natural Language Processing (EMNLP 2010)*, 2010. 1, 68, 69, 70, 72, 87

F. M. Suchanek, G. Kasneci, and G. Weikum. Yago: a core of semantic knowledge. In *WWW*, 2007. DOI: 10.1145/1242572.1242667. 78, 87

F. M. Suchanek, S. Abiteboul, and P. Senellart. Paris: Probabilistic alignment of relations, instances, and schema. *PVLDB*, 2011. DOI: 10.14778/2078331.2078332. 78, 81

Y.-H. Sung, C. Boulis, C. D. Manning, and D. Jurafsky. Regularization, adaptation, and non-independent features improve hidden conditional random fields for phone classification. In *ASRU*, 2007. DOI: 10.1109/ASRU.2007.4430136. 67

C. Sutton and A. Mccallum. *Introduction to Conditional Random Fields for Relational Learning*. MIT Press, 2006. 67

M. Szummer and T. Jaakkola. Partially labeled classification with Markov random walks. In *Proceedings of Advances in Neural Information Processing Systems (NIPS)*, 2001. 36

P. Talukdar and W. Cohen. Scaling graph-based semi supervised learning to large number of labels using count-min sketch. In *Proceedings of the Conference on Artificial Intelligence and Statistics (AISTATS)*, 2014. 55, 56, 57, 58

P. Talukdar, T. Brants, and M. Pereira. A context pattern induction method for named entity extraction. In *Proceedings of the Conference on Computational Natural Language Learning (CoNLL)*, 2006. DOI: 10.3115/1596276.1596303. 72

P. Talukdar, J. Reisinger, M. Pasca, D. Ravichandran, R. Bhagat, and F. Pereira. Weakly-Supervised Acquisition of Labeled Class Instances using Graph Random Walks. In *Proceedings of the Conference on Empirical Methods in Natural Language Processing (EMNLP)*, 2008. DOI: 10.3115/1613715.1613787. 30, 73, 74, 75, 76

P. P. Talukdar and K. Crammer. New regularized algorithms for transductive learning. In *Proceedings of the European Conference on Machine Learning (ECML-PKDD)*, 2009. DOI: 10.1007/978-3-642-04174-7_29. 27, 32, 34, 43, 44, 61, 62, 63, 71, 85, 95

P. P. Talukdar and F. Pereira. Experiments in graph-based semi-supervised learning methods for class-instance acquisition. In *Proceedings of the Annual Meeting of the Association for Computational Linguistics (ACL)*, 2010. 77

P. P. Talukdar, Z. G. Ives, and F. Pereira. Automatically incorporating new sources in keyword search-based data integration. In *Proceedings of the International Conference on Management of data (SIGMOD)*, 2010. DOI: 10.1145/1807167.1807211. 80

106 BIBLIOGRAPHY

P. P. Talukdar, D. Wijaya, and T. Mitchell. Acquiring temporal constraints between relations. In *Proceedings of the International Conference on Information and Knowledge Management (CIKM)*, 2012. DOI: 10.1145/2396761.2396886. 34

M. Thelen and E. Riloff. A bootstrapping method for learning semantic lexicons using extraction pattern contexts. In *Proceedings of the Conference on Empirical Methods in Natural Language Processing (EMNLP)*, 2002. DOI: 10.3115/1118693.1118721. 72

N. Tishby, F. C. Pereira, and W. Bialek. The information bottleneck method. In *Proceedings of the 37th Annual Allerton Conference on Communication, Control, and Computing*, 1999. 38

A. Tomkins. Keynote speech. 2008. 47

K. Toutanova, D. Klein, C. D. Manning, and Y. Singer. Feature-rich part-of-speech tagging with a cyclic dependency network. In *HLT-NAACL '03*, 2003. DOI: 10.3115/1073445.1073478. 68

I. W. Tsang and J. T. Kwok. Large-scale sparsified manifold regularization. In *Proceedings of the Advances in Neural Information Processing Systems (NIPS)*, 2006. 55

K. Tsuda. Propagating distributions on a hypergraph by dual information regularization. In *Proceedings of the International Conference on Machine Learning (ICML)*, 2005. DOI: 10.1145/1102351.1102467. 37

M. Turk and A. Pentland. Eigenfaces for recognition. *Journal of Cognitive Neuroscience*, 3(1), Jan. 1991. DOI: 10.1162/jocn.1991.3.1.71. 3

J. Ugander and L. Backstrom. Balanced label propagation for partitioning massive graphs. In *Proceedings of the International Conference on Web Search and Data Mining (WSDM)*, 2013. DOI: 10.1145/2433396.2433461. 51

B. Van Durme and M. Paşca. Finding cars, goddesses and enzymes: Parametrizable acquisition of labeled instances for open-domain information extraction. In *Proceedings of the AAAI Conference on Artificial Intelligence (AAAI)*, 2008. 72, 73, 74, 75, 76

V. Vapnik. *Statistical Learning Theory*. Wiley Series, 1998. 5

V. V. Vazirani. *Approximation Algorithms*. Springer-Verlag, 2001. 52

L. Velikovich, S. Blair-Goldensohn, K. Hannan, and R. McDonald. The viability of web-derived polarity lexicons. In *Proceedings of the Conference of the North American Chapter of the Association for Computational Linguistics (NAACL)*, 2010. 84

P. O. Vontobel. A generalized blahut-arimoto algorithm. In *Proceedings of the IEEE International Symposium on Information Theory*, page 53. IEEE, 2003. DOI: 10.1109/ISIT.2003.1228067. 37

H. Wallach. *Efficient Training of Conditional Random Fields*, PhD thesis, University of Edinburgh, 2002. 67

F. Wang and C. Zhang. Label propagation through linear neighborhoods. *IEEE Transactions on Knowledge and Data Engineering*, 20(1):55–67, 2008. DOI: 10.1109/TKDE.2007.190672. 9, 16, 26, 86

Z. Wang, Y. Song, and C. Zhang. Knowledge Transfer on Hybrid Graph. In *Proceedings of the International Joint Conferences on Artificial Intelligence (IJCAI)*, 2009. 86

K. Weinberger and L. Saul. Distance metric learning for large margin nearest neighbor classification. *The Journal of Machine Learning Research*, 2009. DOI: 10.1145/1577069.1577078. 20

D. Wijaya, P. P. Talukdar, and T. Mitchell. Pidgin: ontology alignment using web text as interlingua. In *Proceedings of the International Conference on Information & Knowledge Management (CIKM)*, 2013. DOI: 10.1145/2505515.2505559. 1, 78, 79, 80, 81, 82, 83, 87

C. F. J. Wu. On the convergence properties of the EM algorithm. *The Annals of Statistics*, 11(1): 95–103, 1983. DOI: 10.1214/aos/1176346060. 94

D. Zhou, O. Bousquet, T. N. Lal, J. Weston, and B. Schölkopf. Learning with local and global consistency. In *Proceedings of the Advances in Neural Information Processing Systems (NIPS)*, 2004. 29

D. Zhou, J. Huang, and B. Scholkopf. Learning from labeled and unlabeled data on a directed graph. In *Proceedings of the International Conference on Machine Learning (ICML)*, 2005. DOI: 10.1145/1102351.1102482. 85, 87

J. Zhu, J. Kandola, J. Lafferty, and Z. Ghahramani. *Semi-Supervised Learning*, chapter Graph Kernels by Spectral Transforms. MIT Press, 2007. 10, 21, 22, 23, 24, 25, 26

X. Zhu. Semi-supervised learning literature survey. Technical Report 1530, Computer Sciences, University of Wisconsin-Madison, 2005. 5, 13

X. Zhu and Z. Ghahramani. Learning from labeled and unlabeled data with label propagation. Technical report, Carnegie Mellon University, 2002. 38

X. Zhu and A. Goldberg. *Introduction to Semi-Supervised Learning*. Morgan & Claypool Publishers, 2009. DOI: 10.2200/S00196ED1V01Y200906AIM006. 1, 95

X. Zhu, Z. Ghahramani, and J. Lafferty. Semi-supervised learning using gaussian fields and harmonic functions. In *Proceedings of the International Conference on Machine Learning (ICML)*, 2003. 9, 16, 19, 21, 23, 25, 27, 28, 29, 43, 44, 45, 57, 63, 85, 86, 95

V. Zue, S. Seneff, and J. Glass. Speech database development at MIT:TIMIT and beyond. In *Speech Communication*, 1990. DOI: 10.1016/0167-6393(90)90010-7. 63

Authors' Biographies

AMARNAG SUBRAMANYA

Google Research, 1600 Amphitheater Pkwy., Mountain View, CA 94043, USA
Email: asubram@google.com, Web: http://sites.google.com/site/amarsubramanya

Amarnag Subramanya is a Staff Research Scientist in the Natural Language Processing group at Google Research. Amarnag received his Ph.D. (2009) from the University of Washington, Seattle, working under the supervision of Jeff Bilmes. His dissertation focused on improving the performance and scalability of graph-based semi-supervised learning algorithms for problems in natural language, speed, and vision. Amarnag's research interests include machine learning and graphical models. In particular, he is interested in the application of semi-supervised learning to large-scale problems in natural language processing. He was the recipient of the Microsoft Research Graduate fellowship in 2007. He recently co-organized a session on "Semantic Processing" at the National Academy of Engineering's (NAE) Frontiers of Engineering (USFOE) conference.

PARTHA PRATIM TALUKDAR

401 SERC, Indian Institute of Science, Bangalore, India 560012
Email: ppt@serc.iisc.in, Web: http://talukdar.net

Partha Pratim Talukdar is an Assistant Professor in the Supercomputer Education and Research Centre (SERC) at the Indian Institute of Science (IISc), Bangalore. Before that, Partha was a Postdoctoral Fellow in the Machine Learning Department at Carnegie Mellon University, working with Tom Mitchell on the NELL project. Partha received his Ph.D. (2010) in CIS from the University of Pennsylvania, working under the supervision of Fernando Pereira, Zack Ives, and Mark Liberman. Partha is broadly interested in Machine Learning, Natural Language Processing, Data Integration, and Cognitive Neuroscience, with particular interest in large-scale learning and inference over graphs. His past industrial research affiliations include HP Labs, Google Research, and Microsoft Research.

Index

Printed in the United States
by Baker & Taylor Publisher Services